# MEM09002B

## 2015

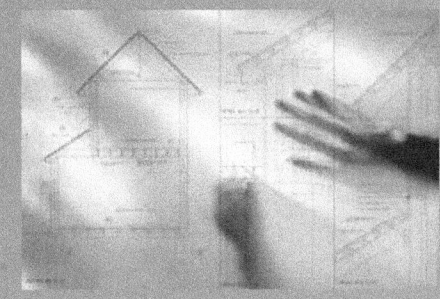

Interpret technical drawing

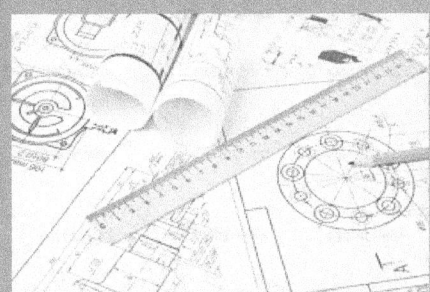

MEM09002B – Interpret technical drawing

First Published January 2013

This work is copyright.   Any inquiries about the use of this material should be directed to the publisher.

Edition 1 – January 2013

Edition 2 – September 2014

Edition 3 – October 2015

MEM09002B – Interpret technical drawing

## Conditions of Use:

Unit Resource Manual

Manufacturing Skills Australia Courses

This Student's Manual has been developed by BlackLine Design for use in the Manufacturing Skills Australia Courses.

Additional resource units can viewed and be ordered at www.acru.com.au

All rights reserved. No part of this publication may be printed or transmitted in any form by any means without the explicit permission of the writer.

**Statutory copyright restrictions apply to this material in digital and hard copy.**

Copyright © BlackLine Design 2015

MEM09002B – Interpret technical drawing

## Feedback:

Your feedback is essential for improving the quality of these manuals.

This learning resource has not been technically edited. Please advise BlackLine Design of any changes, additions, deletions or anything else you believe would improve the quality of this Student Workbook. Don't assume that someone else will do it. Your comments can be made by photocopying the relevant pages and including your comments or suggestions.

Forward your comments to:

>BlackLine Design
>blakline@bigpond.net.au
>Sydney, NSW 2000

## Corporate Licenses

State and National TAFE Colleges and Institutes, and Registered Training Organisations are eligible to purchase corporate licenses.

All licenses are perpetual and allow the licensee to upload the material onto a delivery system (Moodle etc), print the resource in book form and sell or distribute the material to enrolled students within their organisation. The license allows the holder to re-badge the material but must retain acknowledgment to BlackLine Design as the original developer and owner.

MEM09002B – Interpret technical drawing

## Aims of the Competency Unit:

When you have completed this unit of competency you will have developed the knowledge and skills to select the correct and interpret a technical drawing. As a result, you will be able to apply these skills to interpret technical drawing.

## Unit Hours:
36 Hours

## Prerequisites:
None.

# MEM09002B – Interpret technical drawing

## Elements and Performance Criteria

| | | | |
|---|---|---|---|
| 1. | Select correct technical drawing | 1.1 | Drawing is checked and validated against job requirements or equipment. |
| | | 1.2 | Drawing version is checked and validated. |
| 2. | Interpret technical drawing | 2.1 | Components, assemblies or objects are recognised as required. |
| | | 2.2 | Dimensions are identified as appropriate to field of employment. |
| | | 2.3 | Instructions are identified and followed as required. |
| | | 2.4 | Material requirements are identified as required. |
| | | 2.5 | Symbols are recognised in the drawing as appropriate. |

## Required Skills and Knowledge

**Required skills include:**

Required skills include the ability to:
- checking the drawing against job requirements/related equipment in accordance with standard operating procedures
- confirming the drawing version as being current in accordance with standard operating procedures
- where appropriate, obtaining the current version of the drawing in accordance with standard operating procedures
- reading, interpreting information on the drawing, written job instructions, specifications, standard operating procedures, charts, lists and other applicable reference documents
- checking and clarifying task related information
- undertaking numerical operations, geometry and calculations/formulae within the scope of this unit

**Required knowledge includes:**

Competency includes sufficient knowledge to:
- application of AS1100.101 in accordance with standard operating procedures
- relationship between the views contained in the drawing
- objects represented in the drawing
- units of measurement used in the preparation of the drawing
- dimensions of the key features of the objects depicted in the drawing
- understanding of the instructions contained in the drawing
- the actions to be undertaken in response to those instructions
- the materials from which the object(s) are made
- any symbols used in the drawing as described in range statement
- hazard and control measures associated with interpreting technical drawings, including housekeeping
- safe work practices and procedures

# MEM09002B – Interpret technical drawing

## Topic Program:

Unit hour unit and is divided into the following program.

| Topic | Skill Practice Exercise |
|---|---|
| Topic 1 – Engineering Drawings: | MEM09002-RQ-0101 |
| Topic 2 – Drawing Sheets: | MEM09002-RQ-0201 |
| Topic 3 – Line Styles: | MEM09002-RQ-0301 to MEM09002-RQ-0302 |
| Topic 4 – Dimensions: | MEM09002-RQ-0401 |
| Topic 5 – Orthographic Projection: | MEM09002-RQ-0501 to MEM09002-RQ-0502 |
| Topic 6 – Sections: | MEM09002-RQ-0601 to MEM09002-RQ-0602 |
| Topic 7 – Scales: | MEM09002-RQ-0701 to MEM09002-RQ-0703 |
| Topic 8 – Abbreviations, Symbols & Notes: | MEM09002-RQ-0801 to MEM09002-RQ-0802 |
| **Error! Not a valid result for table.** | MEM09002-RQ-0901 to MEM09002-RQ-0902 |
| Topic 10 – Manufacturer's Catalogues: | MEM09002-RQ-1001 |
| Practice Competency Test | MEM09002–PT-01 |

# MEM09002B – Interpret technical drawing

## Contents:

Conditions of Use: ................................................................................................... 3
   *Unit Resource Manual* ............................................................................. *3*
   *Manufacturing Skills Australia Courses* ................................................ *3*
Feedback: ................................................................................................................. 4
Aims of the Competency Unit: ................................................................................ 5
Unit Hours: ............................................................................................................... 5
Prerequisites: ........................................................................................................... 5
Elements and Performance Criteria ...................................................................... 6
Required Skills and Knowledge ............................................................................. 6
Topic Program: ........................................................................................................ 7
Contents: .................................................................................................................. 8

### Topic 1 – Engineering Drawings: ........................................................... 12
Required Skills: ...................................................................................................... 12
Required Knowledge: ............................................................................................ 12
1.1 Introduction: .................................................................................................... 12
1.2 Standards: ........................................................................................................ 13
1.3 Development of the Drawing: ........................................................................ 13
1.4 Types of Drawings: ......................................................................................... 13
1.5 Freehand Sketch: ............................................................................................ 14
1.6 Detail Drawing: ................................................................................................ 15
1.7 Assembly Drawing: ......................................................................................... 15
1.8 Pictorial: ........................................................................................................... 16
   *1.8.1 Isometric:* ............................................................................................ *17*
   *1.8.2 Oblique:* ............................................................................................... *17*
   *1.8.3 Axonometric:* ....................................................................................... *17*
   *1.8.4 Perspective:* ......................................................................................... *18*
1.9 Schematic Diagram: ....................................................................................... 18
Skill Practice Exercises: ....................................................................................... 20

### Topic 2 – Drawing Sheets: ..................................................................... 25
Required Skills: ...................................................................................................... 25
Required Knowledge: ............................................................................................ 25
2.1 Sheet Sizes: ..................................................................................................... 25
2.2 Drawing Sheet Layout: ................................................................................... 25
   *2.2.1 Border:* ................................................................................................. *26*
   *2.2.2 Title Block:* ........................................................................................... *27*
   *2.2.3 Part List:* ............................................................................................... *28*
   *2.2.4 Revision Block:* .................................................................................... *29*
   *2.2.5 Notes & Legend:* .................................................................................. *29*
Skill Practice Exercises: ....................................................................................... 31

### Topic 3 – Line Styles: .............................................................................. 32
Required Skills: ...................................................................................................... 32
Required Knowledge: ............................................................................................ 32
3.1 Line Styles & Conventions: ........................................................................... 32
   *3.1.1 Visible Outlines:* ................................................................................... *32*
   *3.1.2 Hidden Outlines:* .................................................................................. *32*
   *3.1.3 Centrelines:* .......................................................................................... *32*
   *3.1.4 Dimension Lines:* ................................................................................. *33*
   *3.1.5 Extension/Projection Lines:* ................................................................ *33*
   *3.1.6 Break Lines:* ......................................................................................... *33*
   *3.1.7 Cutting Plane Line:* .............................................................................. *33*
   *3.1.8 Phantom Lines:* .................................................................................... *33*
   *3.1.9 Existing or Adjacent Parts:* .................................................................. *34*
   *3.1.10 Typical Example of Line Styles:* ....................................................... *34*
3.2 Precedence of Lines: ...................................................................................... 35
Skill Practice Exercises: ....................................................................................... 36

# MEM09002B – Interpret technical drawing

**Topic 4 – Dimensions:** .................................................................................................. **38**
    Required Skills: ........................................................................................................... 38
    Required Knowledge: .................................................................................................. 38
    4.1 Historical Measurements: .................................................................................... 38
    4.2 Dimensions: .......................................................................................................... 38
    4.3 Features of a Dimension: ..................................................................................... 40
        *4.3.1 Arrowheads:* ............................................................................................ *40*
        *4.3.2 Dimension Line:* ..................................................................................... *40*
        *4.3.3 Dimension Text:* ..................................................................................... *40*
        *4.3.4 Extension:* ............................................................................................... *40*
        *4.3.5 Gap:* ........................................................................................................ *40*
        *4.3.6 Projection Line:* ...................................................................................... *40*
    4.4 Types of Dimensions: ........................................................................................... 40
        *4.4.1 Linear:* ..................................................................................................... *40*
        *4.4.2 Aligned:* ................................................................................................... *41*
        *4.4.3 Angular:* .................................................................................................. *41*
        *4.4.4 Radial:* ..................................................................................................... *41*
        *4.4.5 Leader:* .................................................................................................... *41*
    4.5 Methods for Dimensioning a Drawing: ................................................................ 42
        *4.5.1 Chain Dimensioning:* .............................................................................. *42*
        *4.5.2 Datum Dimensioning:* ............................................................................ *42*
        *4.5.3 Running Dimensioning:* ......................................................................... *42*
        *4.5.4 Combined Dimensioning:* ...................................................................... *42*
    4.6 Dimensions "Not To Scale": ................................................................................. 43
    4.7 Placement of Dimensions: .................................................................................... 43
    4.8 Dimensioning Rules: ............................................................................................. 45
    4.8 Toleranced Dimensions: ....................................................................................... 47
    Skill Practice Exercises: ............................................................................................... 49

**Topic 5 – Orthographic Projection:** ................................................................................ **50**
    Required Skills: ........................................................................................................... 50
    Required Knowledge: .................................................................................................. 50
    5.1 Orthographic Projection: ...................................................................................... 50
        *5.1.1 Basic Views:* ........................................................................................... *51*
        *5.1.2 Developing the Box:* .............................................................................. *51*
    5.2 Projection Systems: .............................................................................................. 51
        *5.2.1 Third Angle Projection:* .......................................................................... *51*
        *5.2.2 First Angle Projection:* ........................................................................... *52*
        *5.2.3 Projection Symbols:* .............................................................................. *52*
    5.3 Number of Views: ................................................................................................. 53
    Skill Practice Exercises: ............................................................................................... 54

**Topic 6 – Sections:** ............................................................................................................ **58**
    Required Skills: ........................................................................................................... 58
    Required Knowledge: .................................................................................................. 58
    6.1 Introduction: .......................................................................................................... 58
    6.2 Cutting Plane: ........................................................................................................ 59
    6.3 Cross Hatching: ..................................................................................................... 59
    6.4 Types of Section: ................................................................................................... 59
        *6.4.1 Full Section:* ............................................................................................ *59*
        *6.4.2 Half Section:* ........................................................................................... *60*
        *6.4.3 Offset Section:* ....................................................................................... *60*
        *6.4.4 Aligned Section:* ..................................................................................... *61*
        *6.4.5 Revolved Section:* .................................................................................. *62*
        *6.4.6 Removed Section:* ................................................................................. *62*
        *6.4.8 Broken Section:* ..................................................................................... *63*
    6.5 Webs, Ribs & Thin Sections: ................................................................................ 63
        *6.5.1 Alternate Method:* .................................................................................. *64*
    6.6 Holes: ..................................................................................................................... 65
    6.7 Sectioned Assembly Views: .................................................................................. 65

  6.8 Parts Not in Section: .................................................................................... 66
  Skill Practice Exercises: .......................................................................................... 67

## Topic 7 – Scales: .............................................................................................. 71
  Required Skills: ........................................................................................................ 71
  Required Knowledge: ............................................................................................... 71
  7.1 Introduction to Scales: ..................................................................................... 71
  7.2 Engineering Scales: .......................................................................................... 71
    *7.2.1 Imperial Scales: ....................................................................................... 72*
    *7.2.2 Metric Scales: .......................................................................................... 72*
  7.3 Reading Scale Rules: ........................................................................................ 72
  7.4 Recommended Scales: ...................................................................................... 74
  7.5 Converting Scaled Dimensions to Full Size: .................................................... 75
  Skill Practice Exercises: .......................................................................................... 76

## Topic 8 – Abbreviations, Symbols & Notes: ................................................. 79
  Required Skills: ........................................................................................................ 79
  Required Knowledge: ............................................................................................... 79
  8.1 Abbreviations & Acronyms: ................................................................................ 79
    *8.1.1 Acronym: .................................................................................................. 79*
    *8.1.2 Initialism: ................................................................................................. 79*
    *8.1.3 Truncation: .............................................................................................. 79*
  8.2 Jargon: ............................................................................................................... 80
  8.3 Symbols: ............................................................................................................ 80
  Skill Practice Exercises: .......................................................................................... 81

## Topic 9 – Reading Drawings: ......................................................................... 83
  Required Skills: ........................................................................................................ 83
  Required Knowledge: ............................................................................................... 83
  9.1 Reading Engineering Drawings: ........................................................................ 83
    *9.1.1 Prerequisites and Definitions: ................................................................ 83*
    *9.1.2 Method of Reading: ................................................................................ 84*
    *9.1.3 Procedure for Reading: .......................................................................... 85*
  9.2 Interpreting Drawings: ....................................................................................... 85
    *9.2.1 Interpreting a drawing in preparation for manufacture: ...................... 85*
    *9.2.2 Interpreting a drawing in preparation for measurement: ..................... 86*
  Skill Practice Exercises: .......................................................................................... 88

## Topic 10 – Manufacturer's Catalogues: ........................................................ 93
  Required Skills: ........................................................................................................ 93
  Required Knowledge: ............................................................................................... 93
  10.1 Introduction: .................................................................................................... 93
  10.2 Reading a Catalogue: ...................................................................................... 93
  10.3 Data Sheet Index: ............................................................................................ 96
  Skill Practice Exercises: ........................................................................................ 124

## Practice Competency Test ............................................................................ 127

## Appendix: ........................................................................................................ 137
    *Appendix 1 – Abbreviations: ............................................................................ 137*
    *Appendix 2 - Common Engineering Drawing Symbols: ................................. 140*
    *Appendix 3 - Welding Symbols: ....................................................................... 141*
    *Appendix 4 – Structural Steel Sections: ......................................................... 142*
    *Appendix 4 – Structural Steel Profiles: ........................................................... 143*
    *Appendix 5 – Pipeline Symbols: ...................................................................... 144*
    *Appendix 6 - Mechanical Symbols: ................................................................. 146*
    *Appendix 7 – Electrical Symbols: .................................................................... 147*
    *Appendix 8 – Electronic Symbols: ................................................................... 150*
  Skill Practice Exercises: ........................................................................................ 152

**Answers:** ................................................................................................... 153

# Topic 1 – Engineering Drawings:

**Required Skills:**
On completion of the session, the participants will be able to:

- Name AS 1100 as the drawing standards setting out the basic principles of technical drawing practice.
- Identify the various types of drawings produced by drawing offices.

**Required Knowledge:**

- AS 1100 Drawing Standards.
- Development procedures for drawings.

## 1.1 Introduction:

Drawings are legal documents and are often referred to in a court of law therefore all drawings MUST be correct BEFORE being released from the office for production or construction.

A drawing is one method of presenting technical communication. Technical communication is an advanced form of communication whereby people of the same trade (profession) can convey messages to one another more accurately and precisely. To achieve this, a technical language (and jargon), which is well standardized, is needed (e.g. botanical names in Horticology and Latin for medical terminology, etc.).

Drawings have been used since the beginning of history for planning and producing art objects, architectural designs and engineering works. Since the Industrial Revolution a system for creating architectural and engineering drawings has evolved. While the pens, pencils, tools and papers for creating drawings have changed, the basic forms for presenting information have stayed the same. People producing technical drawings need to be familiar with the standard ways of presenting design information.

The ability to read and understand information contained on drawings is essential to perform most engineering-related jobs. Engineering drawings are the industry's means of communicating detailed and accurate information on how to manufacture/fabricate, assemble, troubleshoot, repair, and operate a piece of equipment or a system. To understand how to "read" a drawing it is necessary to be familiar with the standard conventions, rules, and basic symbols used on the various types of drawings. Before learning how to read the actual "drawing," an understanding of the information contained in the various non-drawing areas of a print is also necessary.

Draftspersons will inevitably be required to communicate with different people for different reasons as represented in Figure 1.1. In some situations, communications will be sufficiently taken care of by use of plain text. However in other situations, text alone may not suffice and a more specialized form of communication by a technical engineering drawing may prove irreplaceably useful.

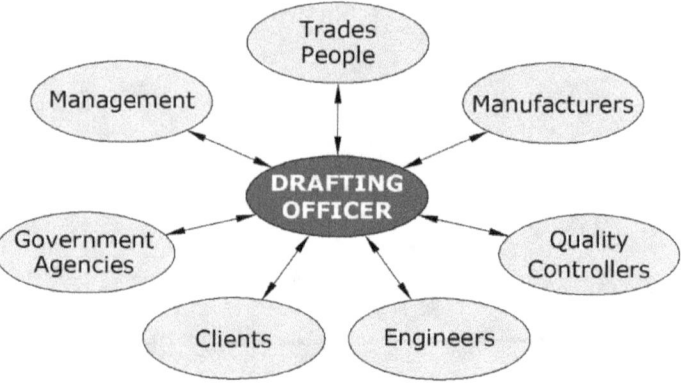

Figure 1.1

## 1.2 Standards:

*"Standardization is the process of formulating and applying rules for an orderly approach to a specific activity for the benefit and with the cooperation of all concerned, and in particular for the promotion of optimum overall economy taking due account of functional conditions and safety requirements."* (International Organization for Standardization).

ISO or International Standards ensure that products and services are safe, reliable and of good quality. For business, they are strategic tools that reduce costs by minimizing waste and errors and increasing productivity. A standard is a document that provides requirements, specifications, guidelines or characteristics that can be used consistently to ensure that the materials, products, processes and services are fit for their purpose.

In Australia, **AS1100 Australian Drawing Standards** sets out the basic principles of technical drawing practice and covers:
- The use of abbreviations.
- Materials, sizes, and layout of drawing sheets.
- The types and minimum thicknesses of lines to be used.
- The requirements for distinct uniform letters, numerals and symbols.
- Recommended scales and their application.
- Methods of projection and of indicating the various views of an object.
- Methods of sectioning.
- Recommendations for dimensioning including size and geometrical tolerancing.
- Conventions used for the representation of components and repetitive features of components.

Some the major industry disciplines include mechanical, automotive, architectural, civil and aeronautical.

## 1.3 Development of the Drawing:

Interpreting information from drawings is an important skill. Engineers and architects must be able to look at a set of plans and mentally picture the shapes of objects. Skilled workers must have the same abilities. Reading a drawing involves a highly developed ability to look at lines on the page and convert the shapes from several pictures to form a three-dimensional mental image. A product basically passes through three main stages; The Concept, then the Drawing Production, and finally manufacture; other stages such as estimating, costing and testing are involved but are supplementary to the main design and production.

Concept Stage

Design/Detail Stage

Manufacture Stage

During the Design/Detail stage, the components and assemblies are constantly being modified and redrawn or edited and passed between designer, detailer and engineer. The drawings once completed and passed, are forwarded to the workshops for manufacture by the trades and production workers.

## 1.4 Types of Drawings:

Drawing is one of the basic forms of visual communication and is used to record objects and actions of everyday life in an easily recognizable manner. There are two major types of drawings: artistic drawings and technical drawings.

Artistic drawings are a form of freehand representation that makes use of pictures to provide a general impression of the object being drawn. There are no hard rules or standards in the preparation of artistic drawings.

Artistic drawings are simply drawn by artists, based more or less on one's talent and skills. Although these drawings are often very attractive, they find very limited use in engineering disciplines.

Technical Drawings are detailed drawings drawn accurately and precisely; they are views of objects that have been prepared with the aid of computer programs or technical drawing/drafting instruments in order to record and transmit technical information. The drawings provide an exact and complete description of things that are to be built or manufactured.

- Technical drawings do not portray the objects the way they directly appear to the eye.
- They make use of many specialized symbols and conventions in order to transmit technical information clearly and exactly.
- To understand and correctly interpret technical drawings, one needs to acquaint oneself with the fundamentals of technical drawing; hence the purpose of this unit of competency.

The presentation of engineering or technical drawings is accomplished through several varying types of drawings including Freehand Sketches, Detail Drawings, Assembly Drawings, Pictorial, Schematic Diagrams and Circuit Diagrams.

## 1.5 Freehand Sketch:

Sketching is the creation of graphic images that are graphical representations or models of objects drawn in proportion but to no particular scale. Freehand sketching is manual sketching with the minimum of tools such as paper and pencil. Technical sketching is the art of creating a technical drawing using freehand without special instruments. Technical sketching requires correct shape or form and more so correct size indication. Generally, drawing tools refer to the materials used as aids when creating drawings and they vary from simple to complex instruments and equipment. However, modern drawing needs have changed dramatically due to the availability of computers. Traditional design and drafting has largely given way to computer design drafting but design sketches will always be needed.

Sketches are helpful in capturing design ideas and trying out different solutions in a fast and inexpensive way; sketches are also useful for recording details of a job "on-site" which will be drawn correctly at a later date in the Drawing Office. Technical sketching is used as aid in conceptualization, spatial visualization and translating imagination into visual models. It could also be used as a means to amplify, clarify and record verbal explanations. Freehand sketching is an economic and effective means of formulating alternate solutions to a given problem so that a choice can be made on the best solution. Preliminary design studies are usually done with freehand sketches because accurate and detailed drawing of design options is expensive and time wasting at the initial stages of a project.

Artistic ability is an asset but anyone can learn to sketch by following basic techniques. Draftspersons and Engineers frequently use special sketching grids which help keep lines straight and in proportion.

Words and notes on sketches must be readable and placed using uppercase characters to assure clarity. Cursive or script writing is never used as it is often unreadable after sketches and memos are duplicated, emailed or faxed to another location. Vertical capital block form letters are preferred.

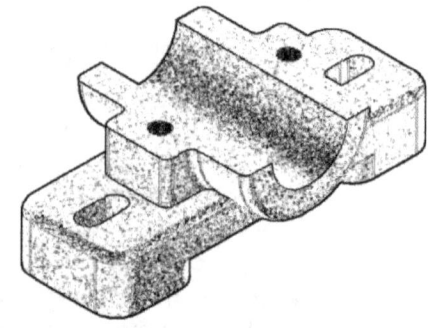

Figure 1.2

# MEM09002B – Interpret technical drawing
## Topic 1 – Engineering Drawings

Figure 1.2 shows a freehand sketch of a Plumber Block Base; the sketch would normally include dimensions and notations but not the shading.

### 1.6 Detail Drawing:
A detail drawing is a print that shows a single component or part. It includes a complete and exact description of the part's shape and dimensions, and how it is made. A complete detail drawing will show in a direct and simple manner the shape, exact size, type of material, finish for each part, tolerance, necessary shop operations, number of parts required, and special notes for the manufacture or treatment after manufacture. A detail drawing is not the same as a detail view. A detail view shows part of a drawing in the same plane and in the same arrangement, but in greater detail to a larger scale than in the principal view.

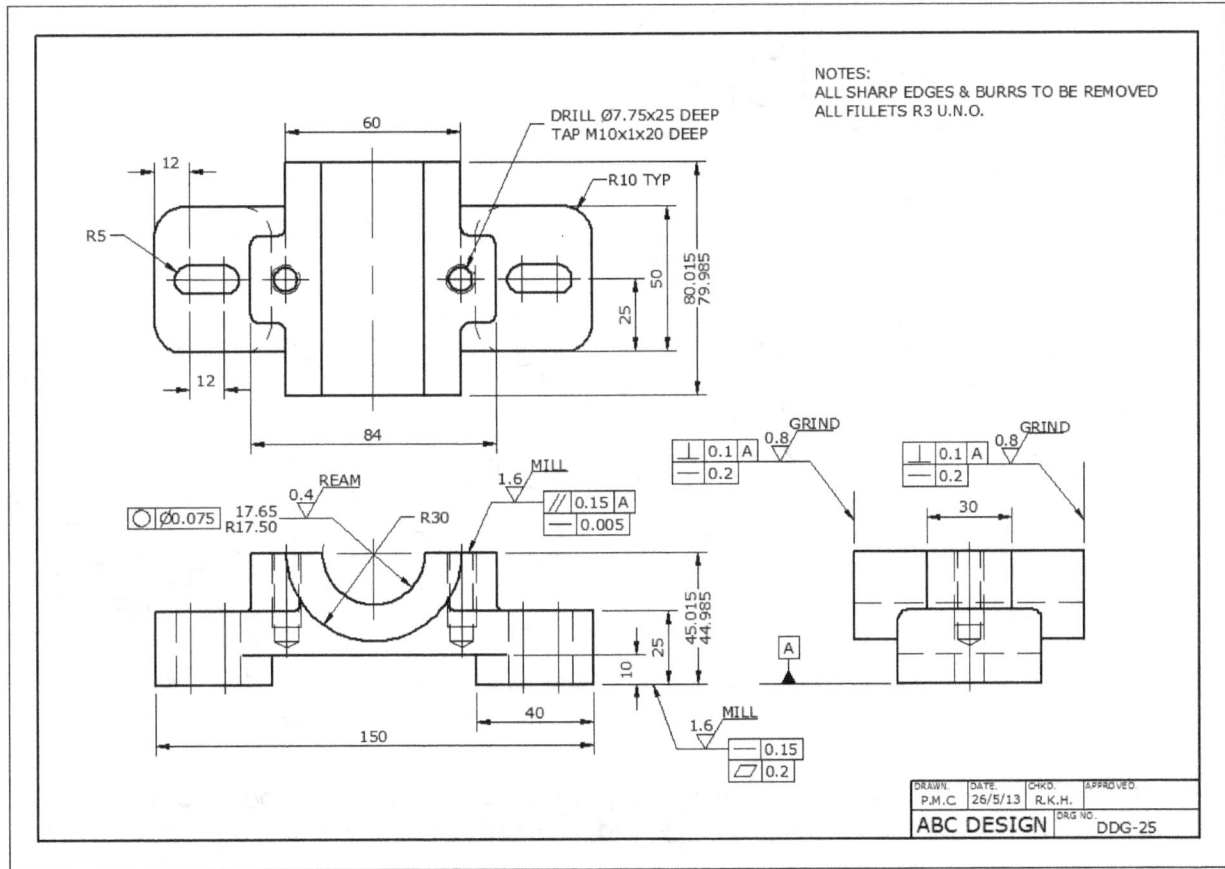

*Figure 1.3*

Figure 1.3 shows a detail drawing of the Plumber Block Base sketched in Figure 1.2. Three views have been provided to describe the shape while all dimensions, surface finish, general and geometric tolerances, and notations have been included on a completed drawing sheet.

### 1.7 Assembly Drawing:
An assembly working-drawing indicates how the individual parts of a machine or mechanism are assembled to make a complete unit. An assembly drawing serves the following purposes:
- Describes the shape of the assembled unit.
- Indicates how the parts of the assembled unit are positioned in relation to each other.
- Identifies each component that forms part of the assembled unit.
- Provides a parts list that describes and lists essential data concerning each part of the assembled unit.

- Provides, when necessary, reference information concerning the physical or functional characteristics of the assembled unit.

Assembly drawings may show one, two or three views to describe the assembled components; they must contain a Parts List (may also be called Material or Cutting List depending on the engineering discipline), cross-referencing (in balloons or circles), and General Notes pertaining to the assembly. The drawings normally show the over dimensions and centre-to-centre distances for specific assemblies.

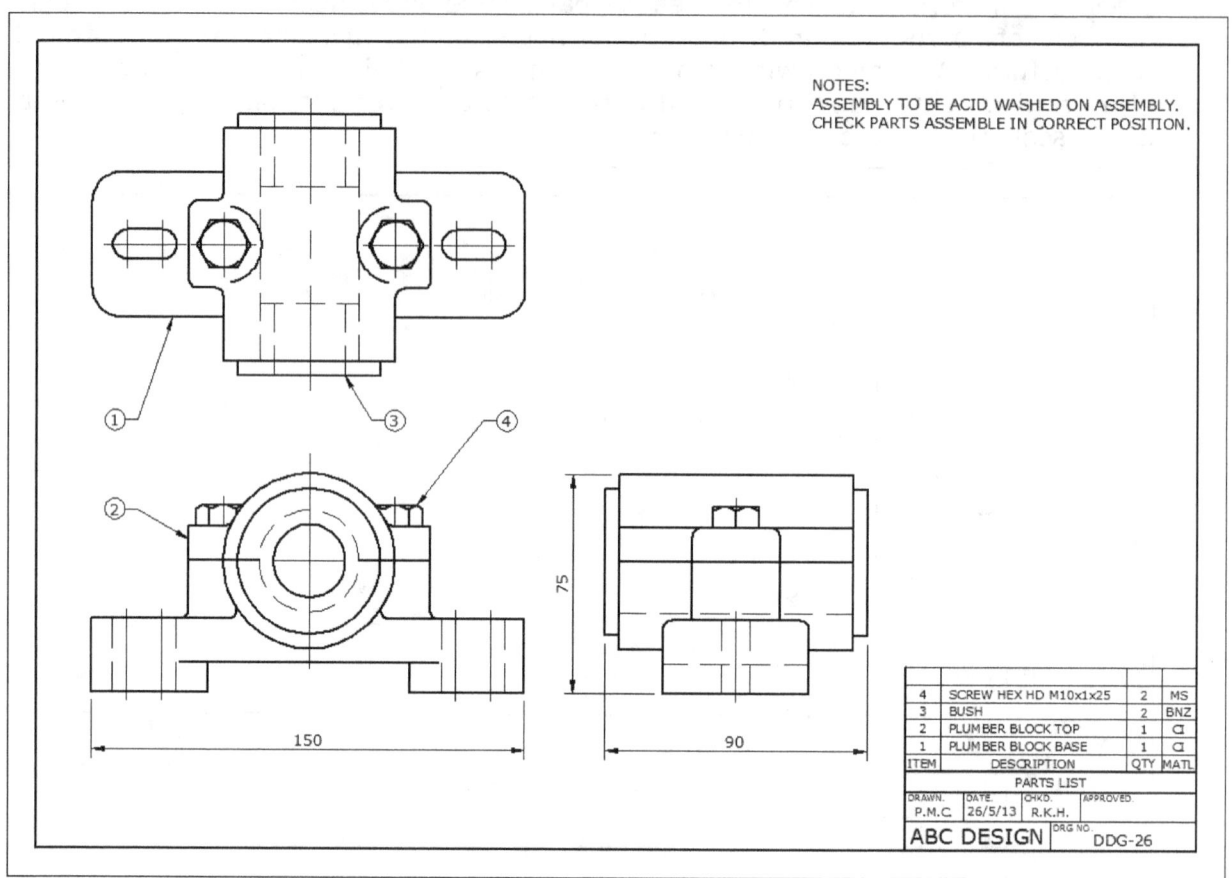

*Figure 1.4*

Figure 1.4 shows a completed assembly drawing with the Plumber Block Base, Plumber Block Top and Bushes drawn in place and secured with the Hexagonal Head Screws.

Most designs are commenced with an assembly drawing and when the concept of the design is finalised, the separate components can be broken out and detailed accordingly.

## 1.8 Pictorial:

Pictorial drawings are wrongly referred to as 3-D drawings. Pictorial drawings represent the shape of an object to show the three principal dimensions (length, width and height); it depicts the way people are used to viewing the object in everyday life but is drawn in 2-D. Characteristics of pictorial drawings are:
- The shapes are easier to visualise and intersections of surfaces can be seen.
- Used for advertising, technical and repair manuals, and for general information.
- Pictorials can distort the lengths of lines and angles at corners; due the distortion factor, pictorial drawings are rarely used for production drawings.
- Pictorial drawings are 2-D drawings where the length along the Z-axis is 0 (zero).

The majority of pictorial drawings are produced as Isometric, Oblique, Axonometric or Perspective drawings.

### 1.8.1 Isometric:

Isometric drawings show three sides in dimensional proportion, but none are shown as a true shape with 90° corners. All the vertical lines are drawn vertically but all horizontal lines are drawn at 30° to the base line. All entities are drawn to scale. Circles and arcs are drawn as ellipses. Isometric is an easy method for presenting 3-D shapes.

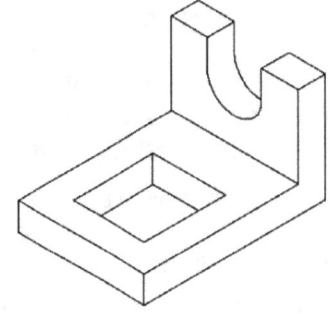

*Figure 1.5*

### 1.8.2 Oblique:

Oblique drawings are also designed to show a three dimensional view of an object. The widths of the object are drawn as horizontal lines, but the depth is drawn back at a 45° angle. Three types of oblique drawings can be used to depict the object, normal, cavalier and cabinet obliques.
- Cavalier drawings display the depth using the full measurement.
- Normal drawings display the depth using ¾ of the measurement.
- Cabinet drawings display the depth using ½ of the measurement.

Circles are easier to draw in oblique as the circles can be drawn using a compass.

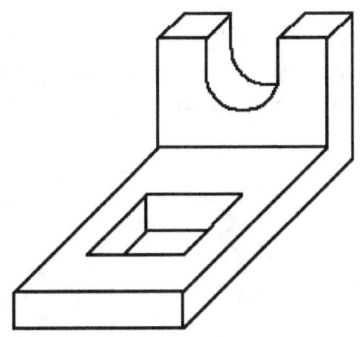 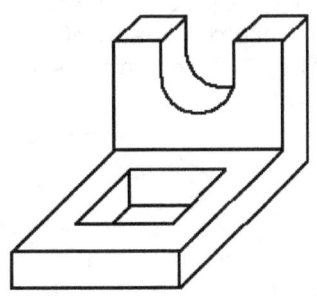 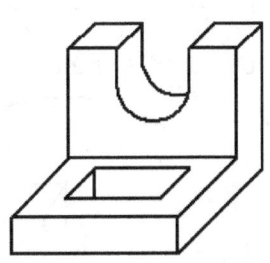

*Figure 1.6 - Cavalier*     *Figure 1.7 - Normal*     *Figure 1.8 - Cabinet*

Of the three images above only Figure 1.7 - Normal appears similar to the real-life object when viewed in oblique, Figure 1.6 - Cavalier appears too elongated while Figure 1.8 - Cabinet is too short or stubby.

### 1.8.3 Axonometric:

In Axonometric drawings, the object's vertical lines are drawn vertically, while the horizontal lines in the width and depth planes are shown at 30°-60° to the horizontal; in other words the Plan or Top View is rotated through 30° or 60°.

Many kitchen manufacturers utilise axonometric in conveying the proposed arrangement of a new kitchen to a client.

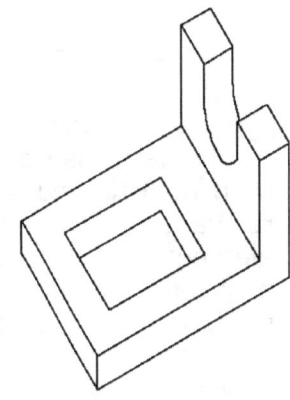

*Figure 1.9*

### 1.8.4 Perspective:

Perspective excels over all other types of projection in the pictorial representation of objects because it more closely approximates the view obtained by the human eye. Geometrically, a photograph is in perspective because the camera captures the same data an eye sees. Perspective is an important tool to designers, architects and engineers however is seldom used apart from architectural applications.

The elements required to produce perspective drawings can become quite daunting; these elements include Picture Planes, Station Points, Left & Right Vanishing Points and Horizon Plane.

Three types of perspective drawings are available, One-point, Two-point and Three point Perspective.

*1.8.4.1 One-Point Perspective:*
In one-point perspective, the object is placed so that two sets of its principal edges are parallel to the Picture Plane, and the third set is perpendicular to the Picture Plane. The third set of parallel lines will converge toward a single vanishing point in perspective.

*1.8.4.2 Two-Point Perspective:*
In two-point perspective, the object is placed so that one set of parallel edges is vertical and has no vanishing point, while the other two sets each have vanishing points. Two-point perspective is the most common type used and is especially suitable for displaying houses and large engineering structures.

*1.8.4.3 Three-Point Perspective:*
In three-point perspective, the object is placed so that none of its principal edges are parallel to the Picture Plane, therefore, each of the three sets of parallel edges will have a separate Vanishing Point. The Picture Plane is assumed approximately perpendicular to the centreline of the cone of rays.

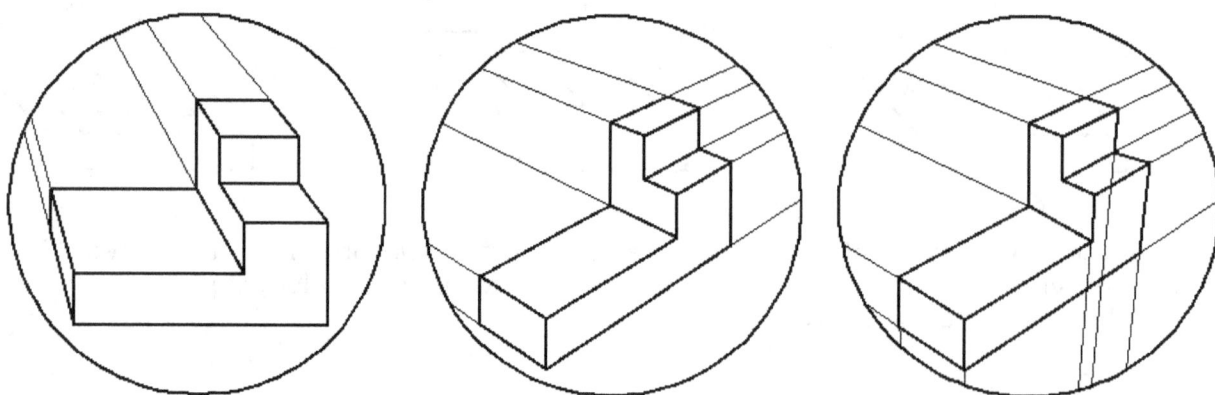

*Figure 1.10 One-Point*  *Figure 1.11 Two-Point*  *Figure 1.12 Three-Point*

### 1.9 Schematic Diagram:

A schematic diagram represents the elements of a system using graphical symbols rather than realistic and detailed drawings. A schematic usually omits all details that are not relevant to the information the schematic is intended to convey, and may add unrealistic elements that aid comprehension. For example, a suburban bus map intended for passengers may represent a bus stop with a dot; the dot doesn't resemble the actual station at all but gives the viewer information without unnecessary visual clutter. A schematic diagram of a chemical process uses symbols to represent the vessels, piping, valves, pumps, and other equipment of the system, emphasizing their interconnection paths and suppressing physical details. In an electronic circuit diagram, the layout of the symbols may not resemble the layout in the physical circuit. In the schematic diagram, the symbolic elements are arranged to be more easily interpreted by the viewer.

# MEM09002B – Interpret technical drawing

## Topic 1 – Engineering Drawings

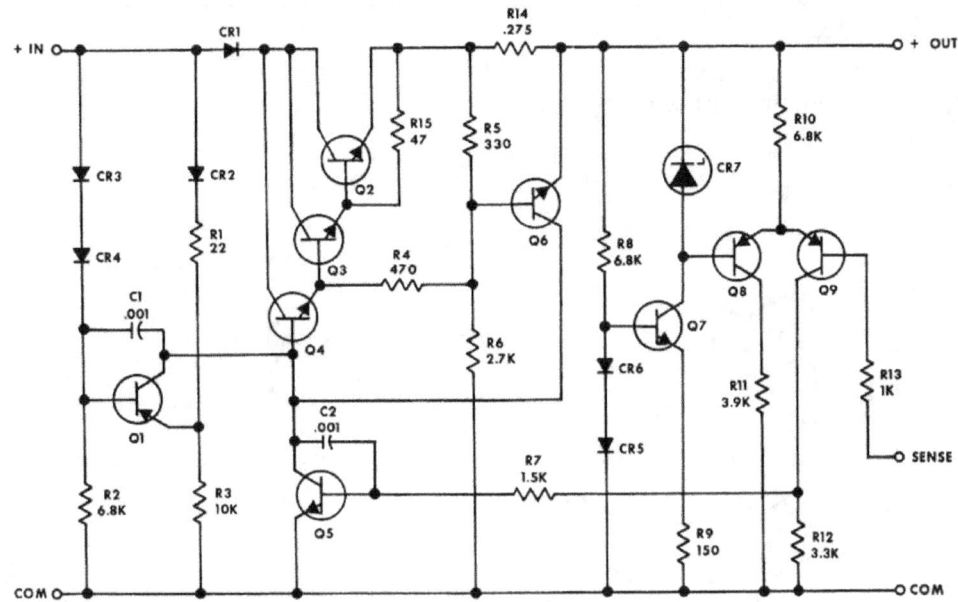

Figure 1.13

MEM09002B – Interpret technical drawing

Topic 1 – Engineering Drawings

## Skill Practice Exercises:

*Skill Practice Exercise MEM09002-RQ-0101:*
Identify the following drawing types:

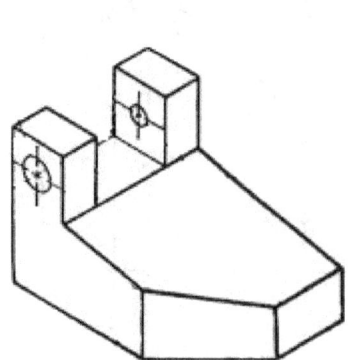

A. _____

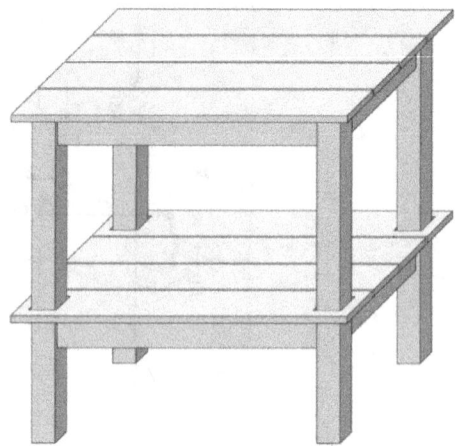

B. _____

C. _____

D. _____

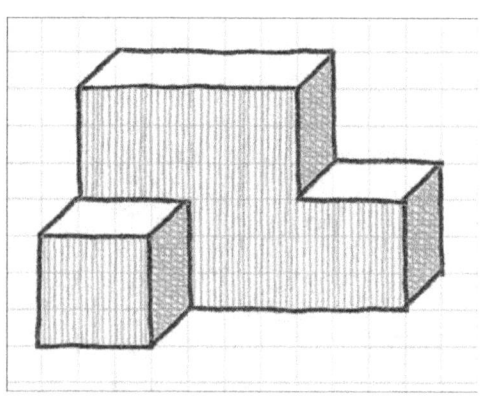

E. _____

F. _____

# MEM09002B – Interpret technical drawing
## Topic 1 – Engineering Drawings

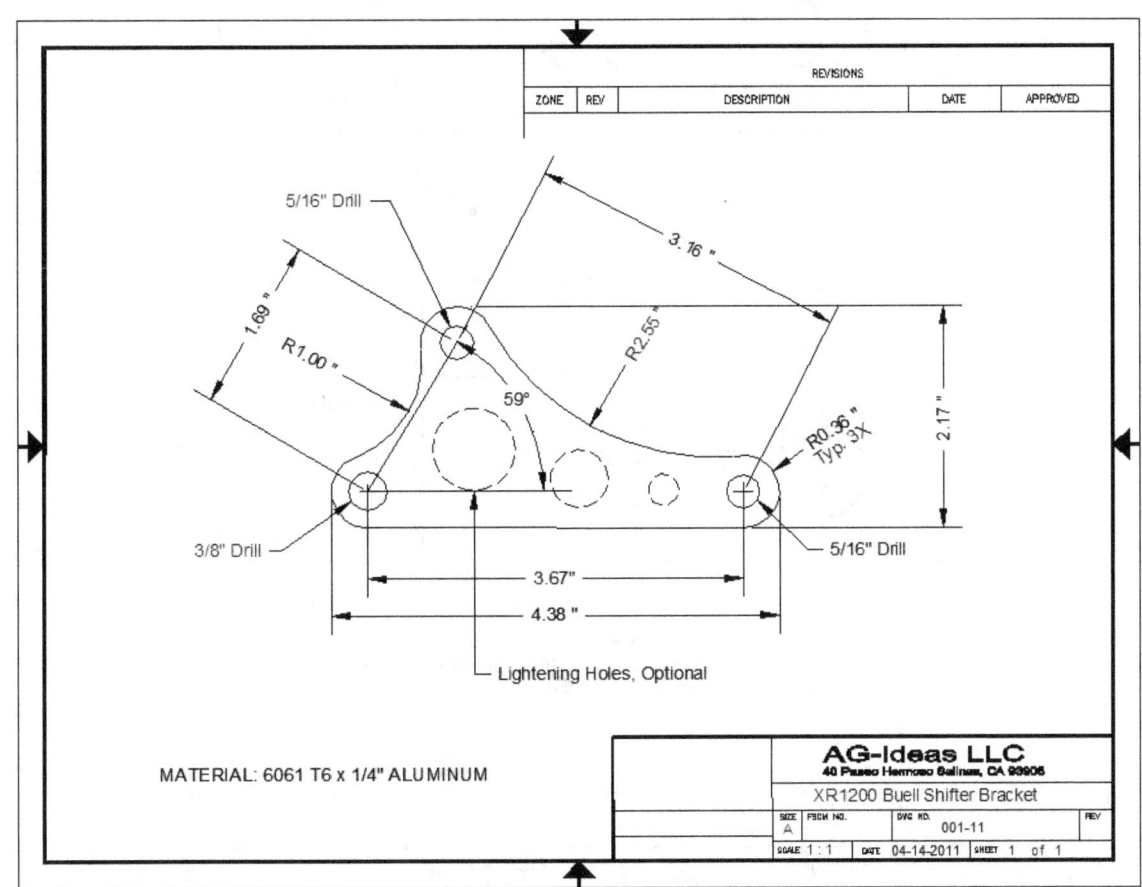

G. _____

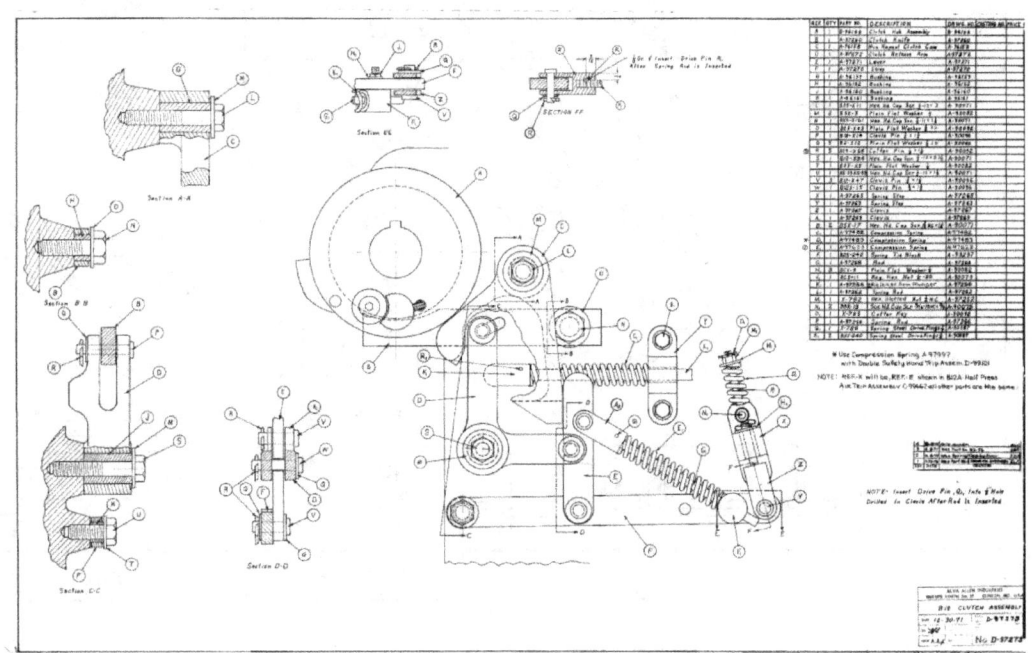

H. _____

# MEM09002B – Interpret technical drawing
## Topic 1 – Engineering Drawings

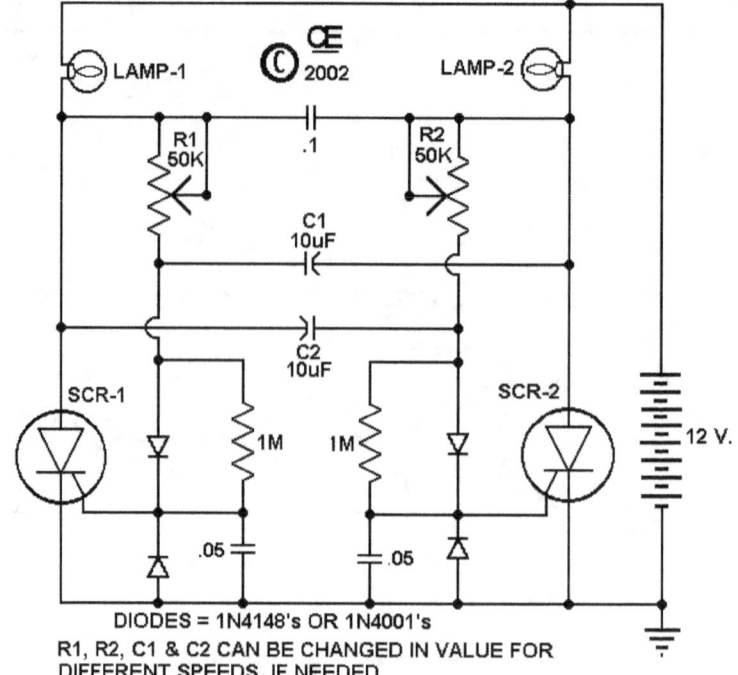

J. _____

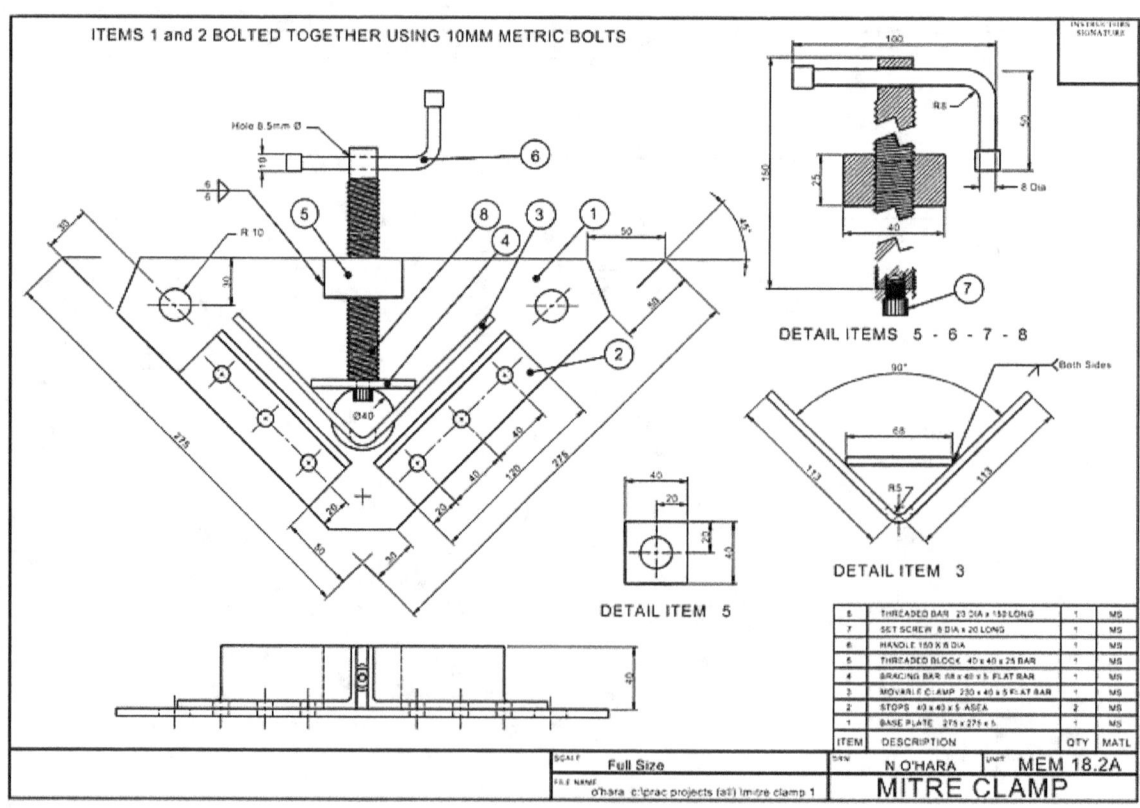

K. _____

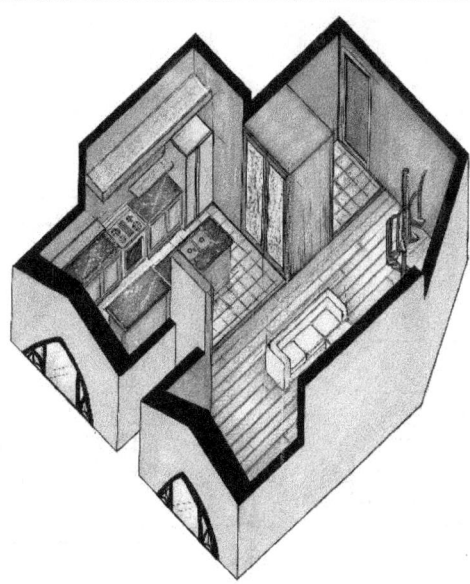

L. _____

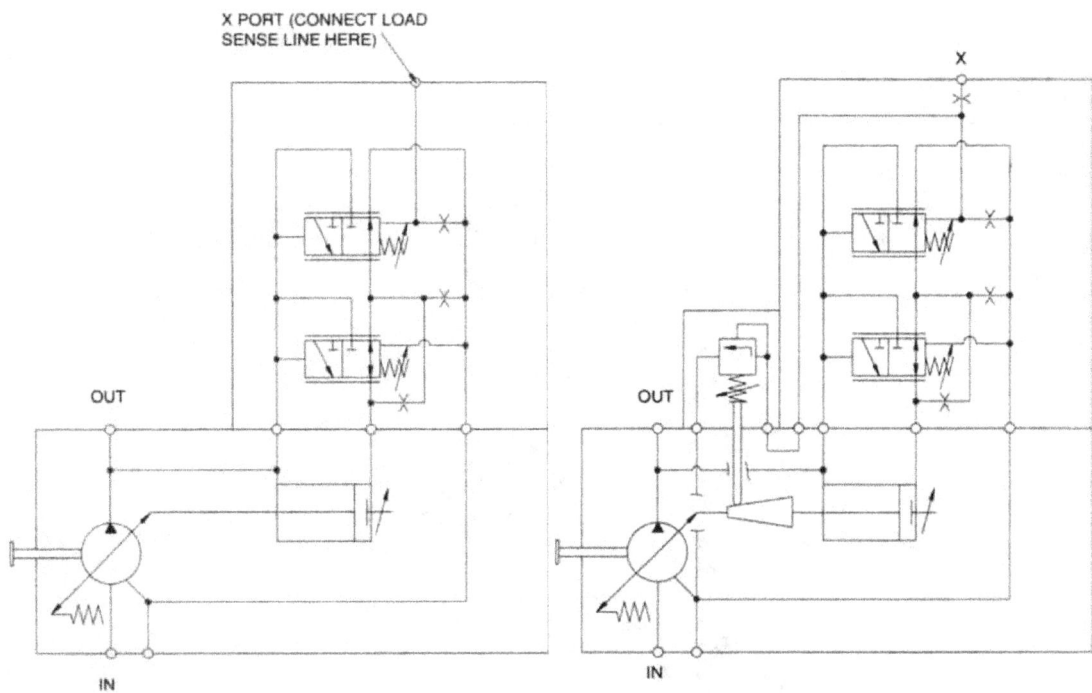

M. _____

# MEM09002B – Interpret technical drawing
## Topic 1 – Engineering Drawings

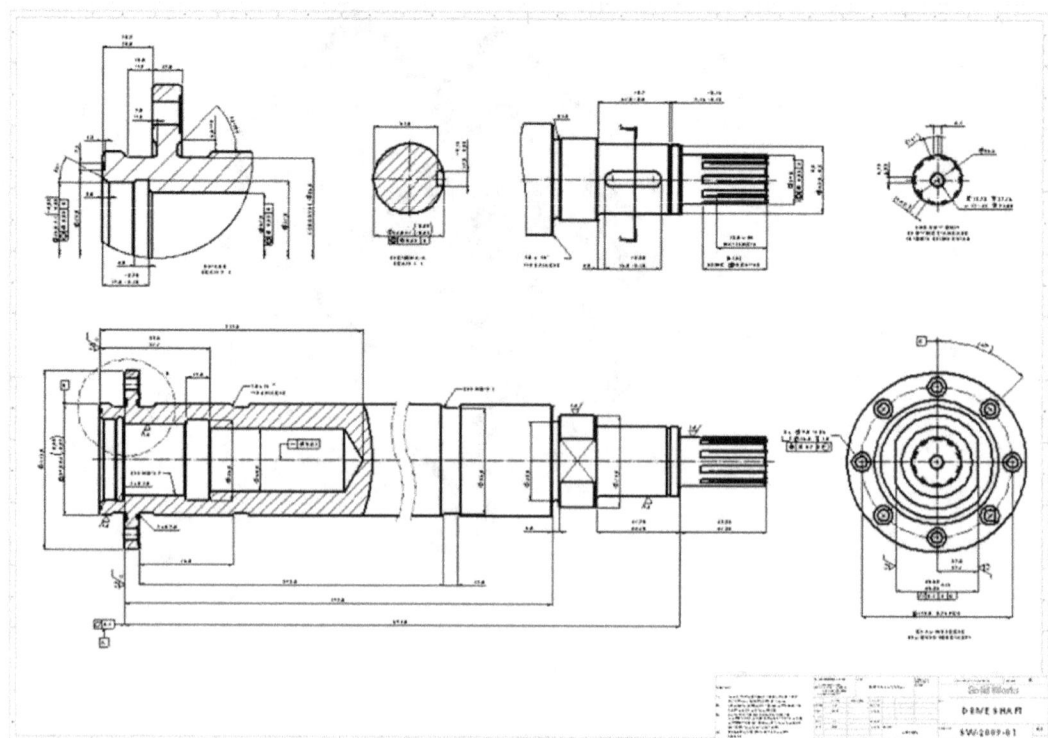

N. _____

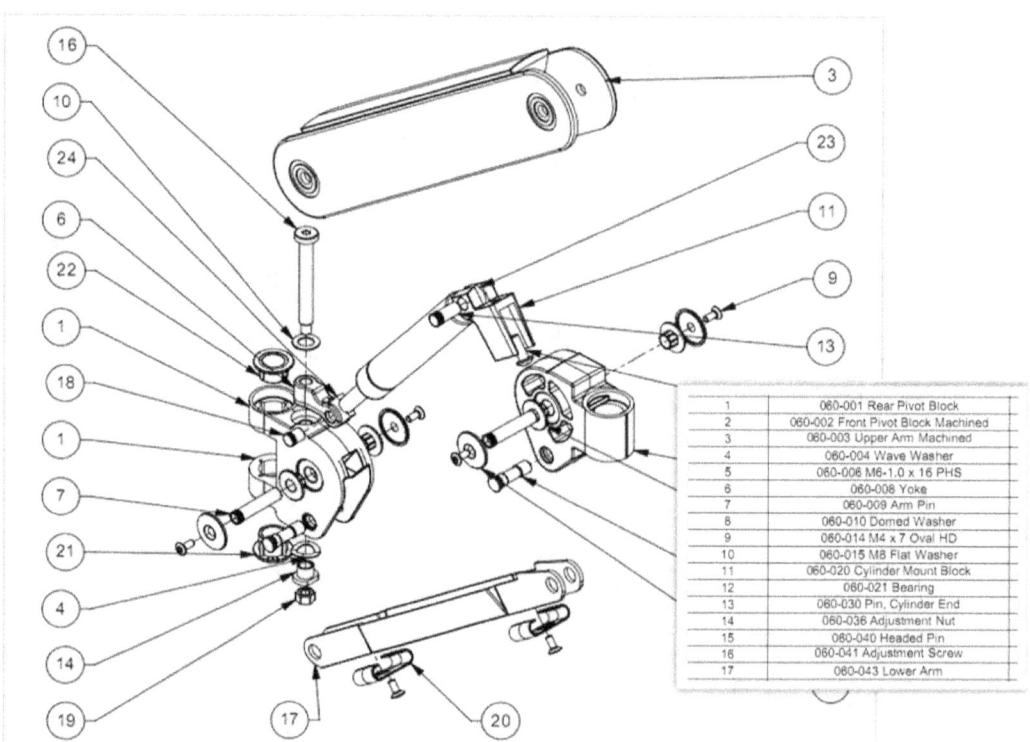

O. _____

Name: _____

# Topic 2 – Drawing Sheets:

**Required Skills:**
- Nominate standard A size sheets.
- Major areas of drawing sheet layouts.

**Required Knowledge:**
- Required data in a Title Block.

## 2.1 Sheet Sizes:
AS 1612-1999 Paper Sizes is similar to the International Standard (ISO 216 - Writing paper and certain classes of printed matter, Trimmed sizes A and B Series) and specifies the paper sizes used with three series of paper sizes: A, B and C series. Series C is primarily used for envelopes. The majority of all drawing offices use the ISO A series.

The ISO A Series of sheet sizes is based on a constant width to length ratio of $1:\sqrt{}$ (rounded to the nearest millimetre). The 'A' size is defined as having an area of one square meter ($1m^2$). Paper weights are expressed in grams per square meter ($gm/m^2$). Each sheet size is exactly half the area of the previous sheet size.

| ISO A Series | Size | |
|---|---|---|
| | Metric (mm) | Imperial (Inches) |
| A0 | 1188 x 840 | 44.81 x 33.11 |
| A1 | 840 x 594 | 33.11 x 23.39 |
| A2 | 594 x 420 | 23.39 x 16.54 |
| A3 | 420 x 297 | 16.54 x 11.69 |
| A4 | 297 x 210 | 11.69 x 8.27 |
| A5 | 210 x 148 | 8.27 x 5.83 |

*Figure 2.1*

## 2.2 Drawing Sheet Layout:
The design and detailing of a component may be produced manually on a drawing board using set squares and compasses, or using a Computer Assisted Design/Drafting (CAD) program. Copies are then printed or plotted on standard size sheets of paper or plastic film and distributed to planning, estimating, ordering sections and finally to the workshops or clients.

The drawing contains five major areas or features:
- Plain or Grid/Zone Border.
- Title Block.
- Parts or Material List.
- Revision Block.
- Notes and Legends.

The first four parts listed above provide important information about the actual drawing. The ability to understand the information contained in these areas is as important as being able to read the drawing itself. Failure to understand these areas can result in improper use or the misinterpretation of the drawing.

# MEM09002B – Interpret technical drawing
## Topic 2 – Drawing Sheets

Because of the extreme variation in format, location of information, and types of information presented on drawings from company to company and site to site, all drawings will not necessarily contain the same format, but will usually be similar in nature. In this unit the terms print, drawing, and diagram are used interchangeably and includes the graphics area, Title Block, grid/border system, Revision Block, Parts/Material List, and the notes and legend.

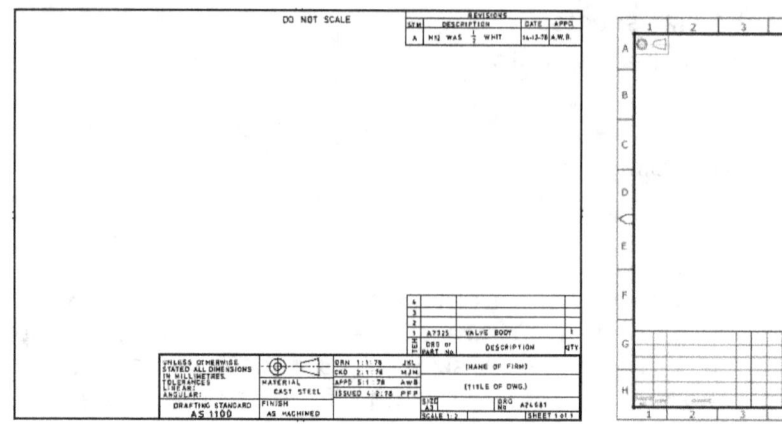
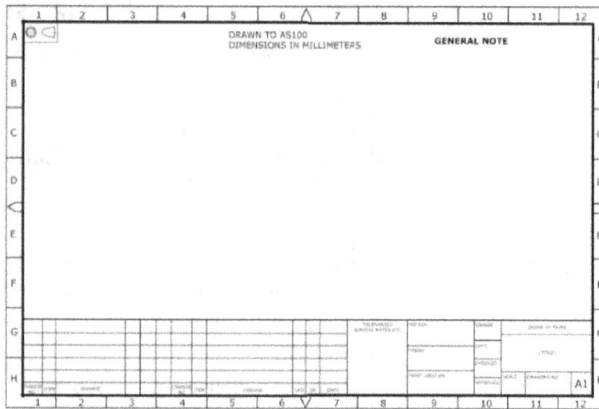

Figure 2.2                                   Figure 2.3

### 2.2.1 Border:
The border is a frame on the four sides and defines the drawing sheet. Borders can consist of a single thick line (0.5 or 0.7 mm) as shown in Figure 2.2 or a grid/zone system.

The sizes for drawing sheets as laid out in AS 1100 is:

| Sheet Size | Nominal Width of Borders | | Dimensions of Drawing Frame |
|---|---|---|---|
| | Sides | Top & Bottom | |
| A0 – 1188 x 840 | 28 | 20 | 1132 x 800 |
| A1 – 840 x 594 | 20 | 14 | 800 x 566 |
| A2 – 594 x 420 | 14 | 10 | 566 x 400 |
| A3 – 420 x 297 | 10 | 7 | 400 x 283 |
| A4 – 297 x 210 | 7 | 5 | 283 x 200 |

The grid or zone border (Figure 2.3) divides the sheet into a grid reference system based on letters and lettered divisions, and allows easy location of features on a large drawing. The grid consists of an outer frame (0.25 mm) and a thicker inner frame (0.7 mm). Where a grid reference system is used, the grid should be laid out in accordance with the following table:

| Detail | Size of Drawing | | | | |
|---|---|---|---|---|---|
| | A0 | A1 | A2 | A3 | A4 |
| Number of vertical zones designated 1, 2, 3 etc. | 16 | 12 | 8 | 8 | 4 |
| Number of horizontal zones designated A, B, C etc. | 12 | 8 | 6 | 6 | 4 |
| Width of margins for grid reference. | 10 | 7 | 7 | 5 | 5 |

*Borders to Suit Plotters :*
Although Australian Standards stipulate the border distances, when designing a border for a sheet to be printed from a computer to a printer, the border distances will have to be adjusted to suit individual plotters because of the grab and carrier head travel

distances. The following table compares 3 plotters with plot areas for standard A3 & A4 sheets.

| Plotter | A3 | A4 |
|---|---|---|
| HP550 | 410.00 x 262.97 | 286.98 x 175.98 |
| OCE5150CA | 408.41 x 275.41 | 285.41 x 188.41 |
| CALCOMP 1026 | 387.91 x 257.31 | 264.91 x 170.31 |

**2.2.2 Title Block:**

The title block of a drawing is usually located in the bottom right corner and contains all the information necessary to identify the drawing and verify its validity. A title block is divided into several areas as illustrated in Figure 2.4, Figure 2.5 and Figure 2.6.

The first area of the title block contains the drawing title, the drawing number, and lists the location, the site, or the vendor. The drawing title and the drawing number are used for identification and filing purposes. Usually the number is unique to the drawing and is comprised of a code that contains information about the drawing such as the site, system, and type of drawing. The drawing number may also contain information such as the sheet number, if the drawing is part of a series, or it may contain the revision level. Drawings are usually filed by their drawing number because the drawing title may be common to several prints or series of prints.

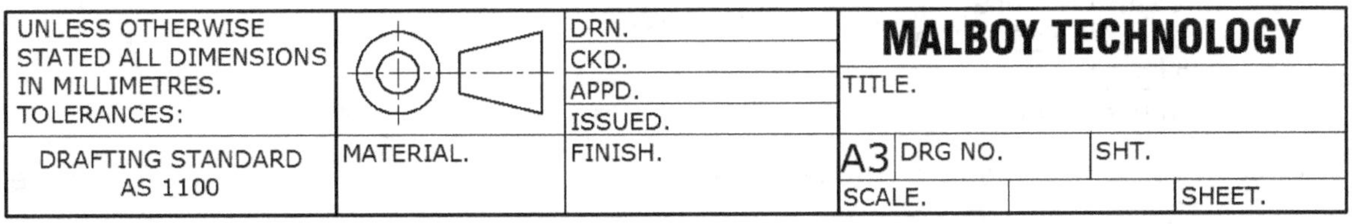

*Figure 2.4*

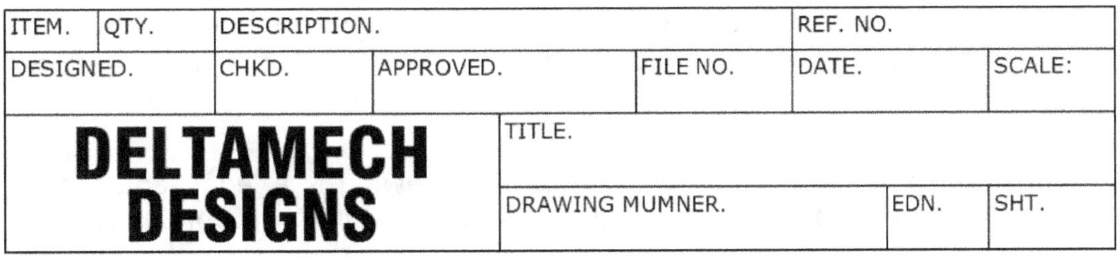

*Figure 2.5*

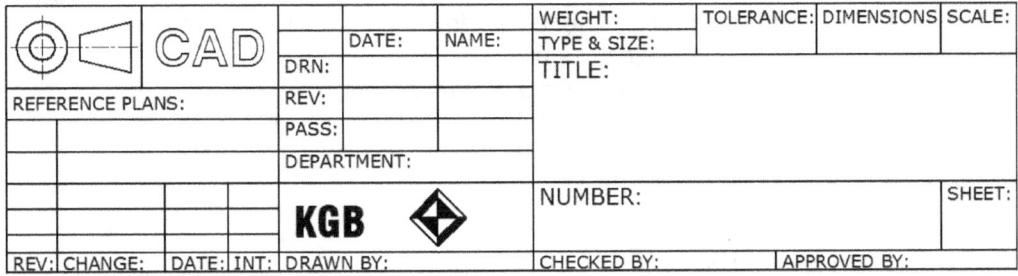

*Figure 2.6*

The second area of the title block contains the signatures and approval dates, which provide information as to when and by whom the component/system was designed and when and by whom the drawing was drafted and verified for final approval. This information can be invaluable in locating further data on the system/component design or operation. These names can also help in the resolution of a discrepancy between the drawing and another source of information.

The third area of the title block is the reference block. The reference block lists other drawings that are related to the system/component, or it can list all the other drawings that are cross-referenced on the drawing, depending on the site's or vendor's conventions. The reference block can be extremely helpful in tracing down additional information on the system or component.

Other information may also be contained in the title block and will vary from site to site and vendor to vendor. Some examples are contract numbers and drawing scale.

Although the three Title Blocks shown above are all totally different in their appearance and layout, they all convey the same or similar data. The main data included in the Title Block is:

*Title:*
The title or name of the project is normally placed using a larger and bolder text than the default size used on the drawing. The title should be brief and as short as possible while containing sufficient information to categorise the part properly and to distinguish it from other similar parts.

The title should list the identifying noun first followed by the modifiers; e.g. FUEL PUMP BRACKET.

*Drawing Number:*
Drawing numbers between different drafting firms and even within companies change as often as the weather. The drawing number is used to locate the drawing in a computerised or physical filing system.

Typical examples are:
- DDG25-113-9600054
- 215/52
- HLK-98-LLK

Drawing numbers can be the same size and font as the default text, or the same as used in the Title.

*Sheet:*
The sheet number normally consists of a numbered consecutively 1, 2, 3,..... The total number of sheets in the set of drawings may be shown but is not a requirement; e.g. 1 of 12.

*Drawn, Checked, Passed, Approved, Examined etc.:*
A record of the personnel who worked on the drawing is a legal requirement. Drawings are legal documents and have been used as evidence in Courts of Law.

*Projection:*
The projection symbol may appear in the Title Block but may be omitted and placed at the top of the drawing sheet. The two methods are Third Angle Project (preferred) or First Angle Projection (only on some foreign drawings). In Australia, Third Angle Projection is the standard method used on drawings.

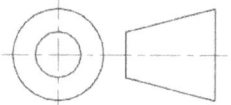 

Third Angle Projection             First Angle Projection

*Name of Company:*
The name of the company that prepared the drawing must be indicated, the inclusion of the company logo is optional. The text should be the same size and font as used for the Title.

### 2.2.3 Part List:
Part Lists are also known as Material Lists or Cutting Lists, or Bill of Materials, depending on the industry. The Part List is a tabulated statement, usually located directly above the Title Block but can be placed on a separate sheet. The table can be simplified into

showing the Item Number, a description, quantity required and material or expanded to provide the stock size of raw material, detail drawing numbers, weight of each piece, specification number etc. A final column can be included for remarks. The term "Part List" (Figure 2.7) applies more accurately in machine-drawing practices. The term "Cutting List" (Figure 2.8) is ordinarily used in structural drawings while the term "Bill of Materials" for architectural drawing. In general, the parts are listed in the order of their importance, with the larger parts first and the standard parts such as screws, pins, etc., at the end.

The blank ruling for a Parts List should not be crowded. Lines should never be spaced closer than 1½ times the text height.

| 5 | SCREW SOC HD CAP M10x1x40 | 8 | MS |
|---|---|---|---|
| 4 | BEARING DEEP GROOVE BALL 6204 | 4 | COMM |
| 3 | SHAFT | 2 | MS |
| 2 | GEARBOX CASING UPPER | 1 | CI |
| 1 | GEARBOX LOWER CASING | 1 | CI |
| ITEM | DESCRIPTION | QTY | MATL |
| | PART LIST | | |

*Figure 2.7*

| 5 | BOLT HEX HD M12 x 1.5 | 50 | AS 2465 | 8 | MS |
|---|---|---|---|---|---|
| 4 | 150 x 12PL | 350 | MIL-ST-501 | 4 | COMM |
| 3 | 75 x 75 x 8EA | 3850 | AS 3679 | 2 | MS |
| 2 | 150UC 37.2 | 2400 | AS 3679 | 1 | CI |
| 1 | 180UB 22.2 | 6750 | AS 3679 | 1 | CI |
| ITEM | DESCRIPTION | LENGTH | SPECIFICATION | QTY | MATL |
| | CUTTING LIST | | | | |

*Figure 2.8*

### 2.2.4 Revision Block:
As changes to a component or system are made, the drawings depicting the component or system must be redrafted and reissued. When a drawing is first issued, it is called revision zero, and the revision block is empty. As each revision is made to the drawing, an entry is placed in the revision block; this entry will provide the revision number, a title or summary of the revision, and the date of the revision. The revision number may also appear at the end of the drawing number or in its own separate block, as shown in Figure 2.9. As the component or system is modified, and the drawing is updated to reflect the changes, the revision number is increased by one, and the revision number in the revision block is changed to indicate the new revision number. For example, if a Revision 2 drawing is modified, the new drawing showing the latest modifications will have the same drawing number, but its revision level will be increased to 3. The old Revision 2 drawing will be filed and maintained in the filing system for historical purposes.

| | | | | | |
|---|---|---|---|---|---|
| D | DETAIL G DELETED | 8-C | LETTER 24/6/13 | D.F.S. | 27/6/13 |
| C | DIM 1250 CHANGED TO 1520 | 10-D | SIG 14785 | W.G.G. | 1/10/12 |
| B | DETAIL G ADDED | 8-C | LETTER 12/5/12 | R.J.B. | 24/5/12 |
| A | ORIGINAL ISSUE | | | | |
| REV | DESCRIPTION | ZONE | AUTH'Y | DRAWN | DATE |
| | CUTTING LIST | | | | |

*Figure 2.9*

### 2.2.5 Notes & Legend:
Drawings are comprised of symbols and lines that represent components or systems. Although a majority of the symbols and lines are self-explanatory or standard (as described in individual competency units), a few unique symbols and conventions must be explained for each drawing. The notes and legends section of a drawing lists and explains any special symbols and conventions used on the drawing. Also listed in the

notes section is any information the designer or draftsperson felt necessary to correctly use or understand the drawing. Because of the importance of understanding all of the symbols and conventions used on a drawing, the notes and legend section must be reviewed before reading a drawing.

Typical notes could include:
- Remove all sharp burrs and edges.
- Component to be x-rayed on completion.
- All machined surfaces to be polished.

MEM09002B – Interpret technical drawing

Topic 2 – Drawing Sheets

## Skill Practice Exercises:

*Skill Practice Exercise MEM09002-RQ-0201:*

1. What is the advantage of using a grid zone border on a drawing sheet?

   _____

2. Give the metric sizes of a standard A2 drawing sheet.

   _____

3. How many vertical zones are shown on a standard A0 sheet?

   _____

4. What is the Australian Standard for paper sizes?

   _____

5. Name the five major features on a drawing sheet.

   _____

6. Title Blocks are normally located in which corner of a drawing?

   _____

7. What method of projection is used in Australia?

   _____

8. If a 100 mm long hexagonal head mild steel M12x1.5 machine screw was to be used in an assembly, show how the description would be shown in a Parts List.

   _____

Name: _____

# Topic 3 – Line Styles:

## Required Skills:
- Types of lines used in engineering drawings.
- Precedence of lines on a drawing.

## Required Knowledge:
- Line construction, widths and/or thicknesses.

## 3.1 Line Styles & Conventions:

Each line on an engineering drawing has a definite meaning and is drawn to a particular construction. The use of different line styles and widths is important in technical drawing as they are used to indicate details and features in a drawing. Line styles make drawings easier to read: for example, solid lines used to show the object will stand out from broken lines used to show hidden information. The correct usage of line styles is essential whether using manual drafting methods or CAD.

Line weight is the thickness of the lines and corresponds to the sheet size being used. AS 1100 incorporates a detailed list of line styles for use in different fields of design including architecture and engineering.

### 3.1.1 Visible Outlines:
Visible outlines are indicated by continuous lines and used to show all edges viewed by the eye when looking at the object. Visible outlines have three possible meanings:
- The intersection of two surfaces.
- Edge view of a surface.
- Contour view of a curved surface.

Visible outlines in CAD have 0.5 or 0.7 lineweights and when drawn in pencil, a heavy or thick line is used. Visible outlines are the predominant line on an engineering drawing and ALWAYS appear in front of other lines where several lines are superimposed.

_____

*Paper Size A4/A3/A2 – 0.5 mm*      *A1/A0 – 0.7 mm*

### 3.1.2 Hidden Outlines:
Hidden lines indicate any surface, edge or feature that cannot be seen because it is on the opposite side of the view from or located inside the object being drawn, e.g. a hole. Hidden lines must join a visible (or another hidden) outline and form "T" and "L" intersections.

Views should be chosen to show features with visible lines where possible. Hidden lines are used to make the object clearer and omitted if not required or can confuse the view. Hidden outlines in CAD have 0.25 or 0.35 lineweights and when drawn in pencil a light or thin line is used.

__  __  __  __  __  __  __  __  __  __  __  __

*Paper Size A4/A3/A2 – 0.25 mm*      *A1/A0 – 0.35 mm*

### 3.1.3 Centrelines:
Centrelines are used to indicate the axes of symmetrical features, circles, holes and paths of motion. The large dashes should cross at the intersection of centrelines and extend uniformly outside the feature for which they are drawn. Centrelines must always start and end with long dashes except for small holes where they can be continuous.

Centrelines in CAD have 0.25 or 0.35 lineweights and when drawn in pencil a light or thin line is used.

——— — ——— — ——— — ——— — ——— — ——— — ——— — ———

*Paper Size A4/A3/A2 – 0.25 mm*     *A1/A0 – 0.35 mm*

### 3.1.4 Dimension Lines:

A dimension line is used to define the measurement of a part feature. Dimension lines consist of a solid line with arrows at both ends and a dimension in the centre.

Dimension in CAD have 0.25 or 0.35 lineweights and when drawn in pencil a light or thin line is used.

—————————————————————————————

*Paper Size A4/A3/A2 – 0.25 mm*     *A1/A0 – 0.35 mm*

### 3.1.5 Extension/Projection Lines:

An extension or projection line is used to visually connect the ends of a dimension line to the relevant feature on the part. Extension lines are drawn perpendicular to the dimension line.

Extension/Projection lines in CAD have 0.25 or 0.35 lineweights and when drawn in pencil a light or thin line is used.

—————————————————————————————

*Paper Size A4/A3/A2 – 0.25 mm*     *A1/A0 – 0.35 mm*

### 3.1.6 Break Lines:

A break line indicates that a portion of the item is not shown on the drawing and is necessary for reasons of space or drawing clarity. Break lines can be straight or curved

Break lines in CAD have 0.25 or 0.35 lineweights and when drawn in pencil a light or thin line is used.

—————————————————————————————

*Paper Size A4/A3/A2 – 0.25 mm*     *A1/A0 – 0.35 mm*

### 3.1.7 Cutting Plane Line:

The cutting plane line indicates the location of a view from where the Sectional View is taken. The arrowheads indicate the direction in which the cutaway object is viewed.

Cutting plane lines in CAD have a 0.5 line indicating the cutting plane and an arrow on a 0.25 or 0.35 lineweights and when drawn in pencil a thick and thin lines are used.

*Paper Size A4/A3/A2 – 0.25 mm*     *A1/A0 – 0.35 mm*

### 3.1.8 Phantom Lines:

Phantom lines are used most frequently to indicate an alternate position of a moving part. The part in one position is drawn in full lines, while in the alternate position it is drawn in phantom lines. Phantom lines are also used to indicate a break when the nature of the object makes the use of the conventional type of break unfeasible

Phantom lines in CAD have 0.25 or 0.35 lineweights and when drawn in pencil a light or thin line is used.

———— — ———— — ———— — ———— — ————

*Paper Size A4/A3/A2 – 0.25 mm*     *A1/A0 – 0.35 mm*

# MEM09002B – Interpret technical drawing
## Topic 3 – Line Styles

**3.1.9 Existing or Adjacent Parts.**

Existing or adjacent parts represent any structure or part immediately in the vicinity of the object.

Existing or adjacent part lines in CAD have 0.25 or 0.35 lineweights and when drawn in pencil a light or thin line is used.

*Paper Size A4/A3/A2 – 0.25 mm          A1/A0 – 0.35 mm*

**3.1.10 Typical Example of Line Styles:**

The drawing shown in Figure 4.1 shows the outline of a Plate with a series of drilled holes. For clarity, the dimensions are shown as smaller text while the larger text indicates the type of line.

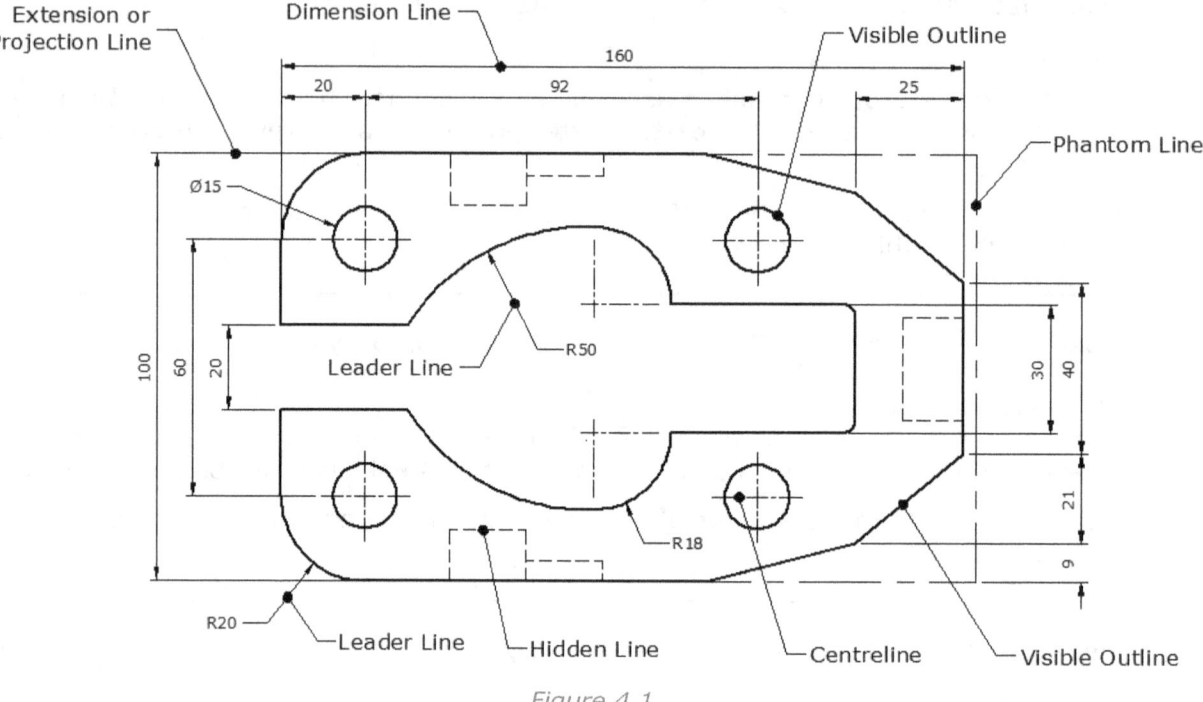

*Figure 4.1*

## 3.2 Precedence of Lines:

On any drawing two or more line styles are going to coincide. Hidden portions of the object may project to coincide with visible portions while centrelines may occur where there is a visible or hidden outline of some feature.

Since the physical features of the object must be represented, full and dashed lines take precedence over all other lines. AS visible outlines are more prominent then dashed lines, they take precedence; visible outlines can cover a hidden outline line but a hidden outline cannot cover a visible outline.

When centrelines and cutting plane lines coincide, the one that is more important takes precedence over the other. Break lines should be placed so they do not spoil the readability of the drawing. Dimension and projection lines must always be placed so as not to coincide with other lines on the drawing.

The precedence of lines are:
- Visible outline
- Hidden outline
- Centreline or Cutting plane line
- Break line
- Dimension & Projection line
- Cross hatching

MEM09002B – Interpret technical drawing

## Topic 3 – Line Styles

### Skill Practice Exercises:

*Skill Practice Exercise MEM09002-RQ-0301:*
Referring to the following images, name the indicated line style in the space provided.

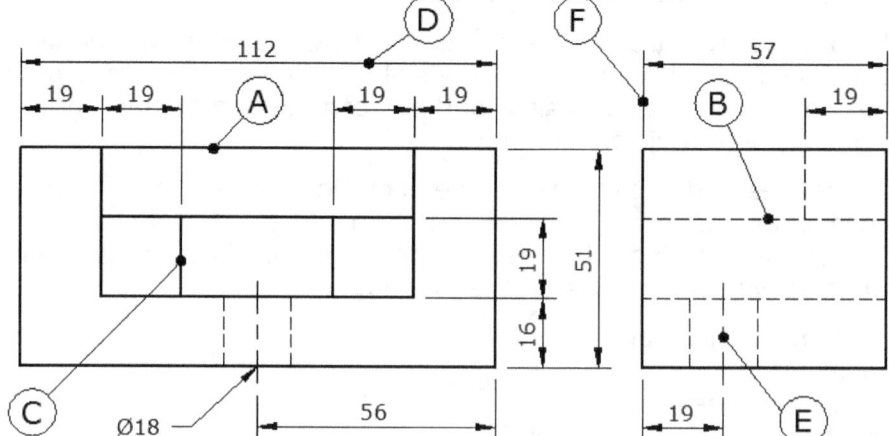

A. _____    B. _____

C. _____    D. _____

E. _____    F. _____

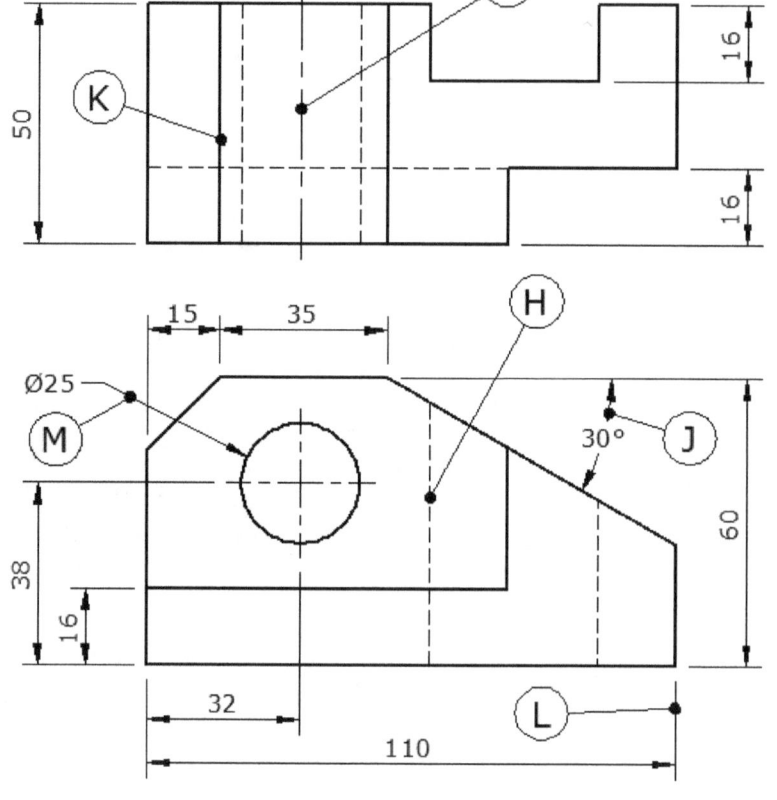

G. _____    H. _____

J. _____    K. _____

L. _____    M. _____

**BlackLine Design**
4th October 2015 – Version 3

# MEM09002B – Interpret technical drawing
## Topic 3 – Line Styles

*Skill Practice Exercise MEM09002-RQ-0302:*
Referring to the following images, name the indicated line style in the space provided.

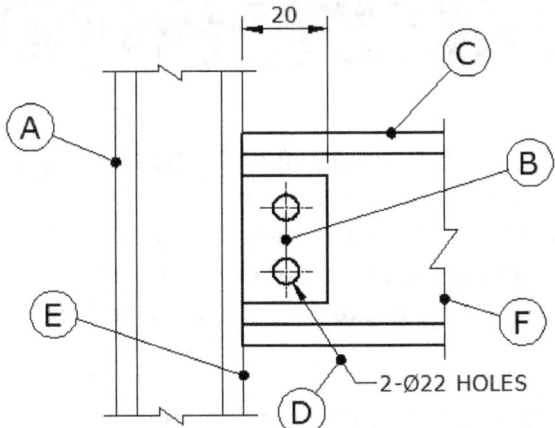

A. _____     B. _____

C. _____     D. _____

E. _____     F. _____

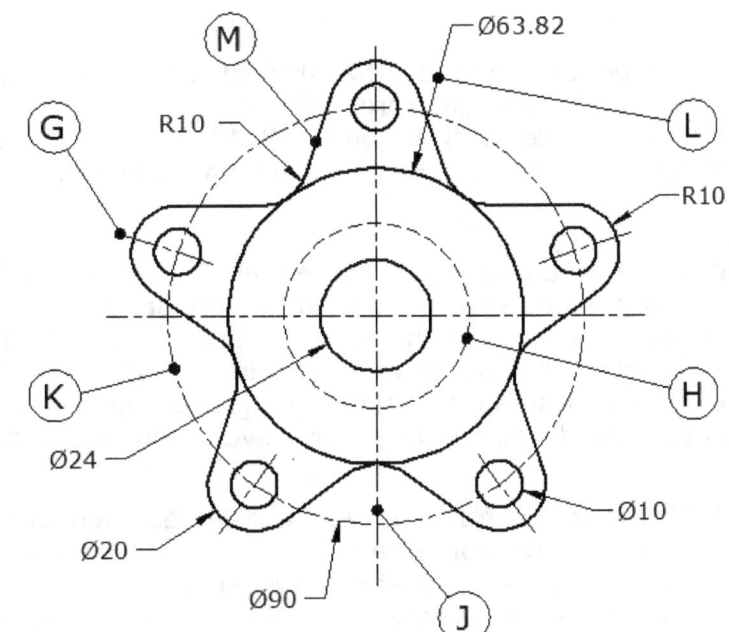

G. _____     H. _____

J. _____     K. _____

L. _____     M. _____

Name: _____

# Topic 4 – Dimensions:

## Required Skills:
- Read and interpret dimensions of objects on a drawing.
- Identify features of a dimension.
- Distinguish the different dimensioning types.
- Name the various methods of dimensioning.
- Identify Unilateral, Bilateral and Direct or Limit of Size toleranced dimensions.

## Required Knowledge:
- AS 1100 rules governing dimensioning.
- Units of measurement used in the preparation of the drawing.
- Dimensions of the key features of the objects depicted in the drawing.

### *Definition:*
*A dimension is a numerical value expressed in appropriate units of measurement and used to define the size, location, orientation, form or other geometric characteristics of a part.*

**NOTE:**
This topic is intended to introduce the student to the concept of dimensions and demonstrate how to read and interpret the data. The application of dimensions to engineering drawings is covered in units MEM09003B - Prepare basic engineering drawing and MEM09005B - Perform Basic Engineering Drafting.

### 4.1 Historical Measurements:
In 1791 France adopted the *meter* which equals 39.37 inches (1" = 25.4 mm) from which the decimalized metric system evolved. At the same time England was setting up a more accurate determination of the *yard* (1 yard = 36 inches) which was defined in 1824 by an act of Parliament. A foot was one third of a yard, and an inch was one thirty-sixth of a yard. From these specifications, graduated rulers, scales and many types of measuring devices were developed to achieve even more accuracy of measurement and inspection.

Until the mid-1900's, common fractions were considered adequate for dimensions, but as the designs became more complicated, and as it became necessary to have interchangeable parts in order to support mass production, more accurate specifications were required and it became necessary to turn to the decimal-inch system. In 1966 Australia commenced the change to the metric system by changing the currency from the British style pound, shilling and pence to the dollar but it until the 1970's that the major changes occurred. 1974 saw the wholesale changes to most Australian industry. Metrication was completed in Australia in 1988.

The current rapid growth of worldwide science and business has fostered an international system of SI units based on the meter and suitable for measurements in engineering. The six basic units of measurement are meter (length), kilogram (mass), second (time), ampere (electric current), degree Celsius (temperature) and candela (luminous intensity or light).

### 4.2 Dimensions:
The growth of the Industrial Revolution and machine manufacturing, the designing and production process were mainly allied in the same building or even by the same person. It was up to the tradesperson to make the objects fit by scaling objects off a drawing.

## Topic 4 - Dimensions

The need for interchanging of parts requires that the drawing be detailed (or fully dimensioned) so parts can be manufactured in different places (or countries) and the mating parts will fit correctly when brought together or assembled. The increased need for precision manufacturing and the necessity to control sizes for interchanging has shifted the responsibility from the tradesperson to the designer and draftsperson.

After the shape of an object has been described by orthographic (or pictorial) views, the value of the drawing for the construction of the object depends upon dimensions and notes that describe the size. In general, the description of shape and size together gives complete information for producing the object represented. The dimensions put on the drawing are not necessarily those used in making the drawing but are those required for the proper functioning of the part after assembly, selected so as to be readily usable by the workers who are to make the piece. Before dimensioning the drawing, draftspersons study the components and understand their functional requirements; then put themselves in the place of the tradesperson and mentally construct the object to discover which dimensions would best give the information.

The basic factors in dimensioning practice are:

1. Lines and Symbols.
   The first requisite is a thorough knowledge of the elements used for dimensions and notes and of the weight and spacing of the lines on the drawing. These lines, symbols, and techniques are the "tools " for clear, concise representation of size.

2. Selection of Distances.
   The most important consideration for the ultimate operation of a machine and the proper working of the individual parts is the selection of distances to be given. This selection is based upon the functional requirements, the "breakdown" of the part into its geometric elements, and the requirements of the shop for production.

3. Placement of Dimensions.
   After the distances to be given have been selected, the next step is the actual placement of the dimensions showing these distances on the drawing. The dimensions should be placed in an orderly arrangement that is easy to read and in positions where they can be readily found.

4. Dimensioning Standard Features.
   These include angles, chamfers, standard notes and specifications of holes, spherical shapes, round-end shapes, tapers, and others for which, through long usage and study, dimensioning practice has been standardized.

5. Precision and Tolerance.
   The ultimate operation of any device depends upon the proper interrelationship of the various parts so that they operate as planned. In quantity production, each part must meet standards of size and position to assure assembly and proper functioning. Through the dimensioning of the individual parts, the limits of size are controlled.

6. Production Methods.
   The method of manufacturing (casting, forging, etc.) affects the detailed information given on the drawing proper, and in notes and specifications. The operations of various shops must be known, in order to give concise information.

## 4.3 Features of a Dimension:

A dimension consists of projection lines, dimension line, dimension text, arrows, projection line extension and gap from the visible outline to the start of the projection line.

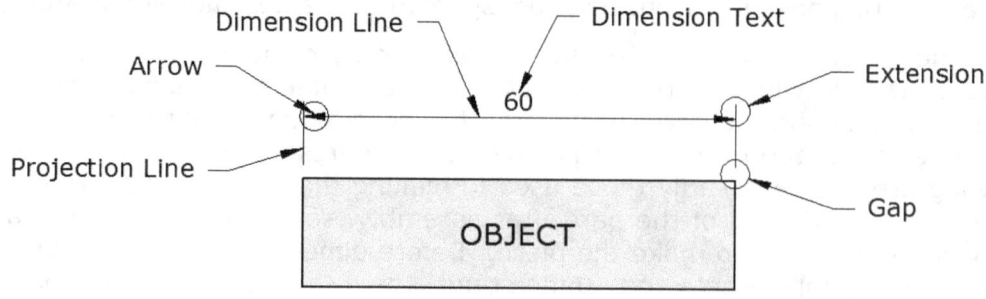

*Figure 4.2*

### 4.3.1 Arrowheads:

Arrowheads indicate the extent of the dimensions and must tough the projection line; they are uniform in style and size. Generally arrows are 3 mm long by 1 mm wide.

### 4.3.2 Dimension Line:

The dimension line is a thin continuous line terminating at the arrowheads. The first dimension line should be placed approximately 15 mm from the nearest outline; other parallel dimensions are equally spaced at approximately 7 mm spacing. Dimension lines are perpendicular to the projection lines.

### 4.3.3 Dimension Text:

The height of the dimension text is the same height as the text used for notes. Text can be placed using the Unidirectional or Aligned Systems.

### 4.3.4 Extension:

To prevent the dimensions appearing similar to the drawing, the projection line is extended past the dimension line.

### 4.3.5 Gap:

A gap between the feature within the view and the projection line is provided to prevent the person reading the drawing from misinterpreting the information; it shows where the dimension originates and is not part of the detail.

### 4.3.6 Projection Line:

The projection line is a thin continuous line that extends from a point inside the view to which the dimension refers. Except in special circumstances, all projection lines are perpendicular to the feature being dimensioned.

## 4.4 Types of Dimensions:

Several types of dimensions may appear on an engineering drawing, these include linear, aligned, angular, radial and leader.

### 4.4.1 Linear:

Linear dimensions measure the distance between two points in a vertical or horizontal direction and placed on a drawing using the Aligned System or the Unidirectional System. One method only is used on a drawing and are never mixed. The majority of drawing offices in Australia use the Aligned System.

*4.4.1.1 Aligned System:*

The Aligned System refers to the practice of placing all dimension values and notes either vertically or horizontally, depending on the placement of the dimension with the values and notes being read from the right and bottom sides of the drawing.

*4.4.1.2 Unidirectional System:*
The Unidirectional System refers to the practice of placing all dimension values and notes horizontally within the drawings, which are read from the bottom of the sheet.

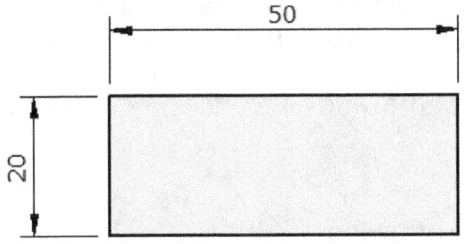
Figure 4.3 - Aligned System

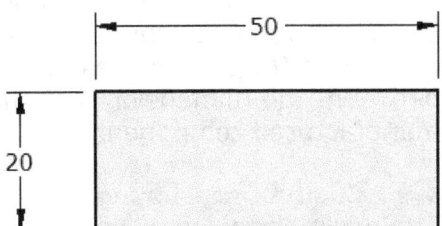
Figure 4.4 – Unidirectional System

**4.4.2 Aligned:**
An aligned dimension is a dimension measured at an angle; i.e. not vertical or horizontal.

**4.4.3 Angular:**
Angular dimensions are formed by an angle distance between two lines and is given in degrees (e.g. 60°). In the civil and architectural industries, the angle may be given in degrees, minutes and seconds (e.g. 125° 35′ 45″).

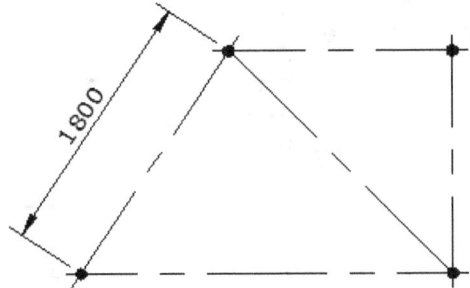
Figure 4.5 – Aligned Dimension

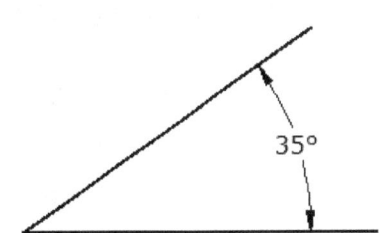
Figure 4.6 – Angular Dimension

**4.4.4 Radial:**
A radial dimension specifies the diameter of circles and the radius of arcs. The arrow touches the circle/arc and points towards the centre if the dimension is placed outside the feature. The dimension can be placed either internally or externally. Radial dimensions are identified by the diameter symbol (Ø) or the letter 'R' to denote radius.

**4.4.5 Leader:**
Leader dimensions are straight lines leading from an explanatory note to the feature on the drawing to which the note applies. The arrowhead always points to the feature and NOT the note. The note end of the leader ends with a short horizontal line, the length is normally the same height as the text, at the mid-point of the note.

Leaders are normally drawn at an angle to contrast with the lines forming the drawing which are mainly horizontal or vertical. Australian Standards allow the use of a "dot" instead of an arrow if the clarity of the view is improved.

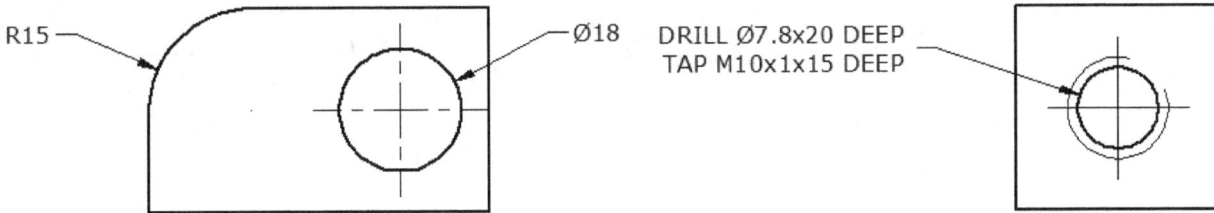
Figure 4.7 – Radial Dimension    Figure 4.8 - Leader

## 4.5 Methods for Dimensioning a Drawing:

Dimensioning a drawing is about adding dimensions, notes and lines to a drawing. There are four methods for dimensioning drawings; chain, datum, running and combination. Some methods of dimensioning a drawing can create accumulating errors, especially when tolerances are added together. An accumulating error is when tolerance errors of different features have been added together. A tolerance error is the difference between the dimension on the engineering drawing and the measurement of the manufactured component.

### 4.5.1 Chain Dimensioning:

Chain dimensioning uses minimal space on a drawing, however, it accumulates an error. Due to the high risk of errors, chain dimension is only used when the function of the component is not affected by the errors.

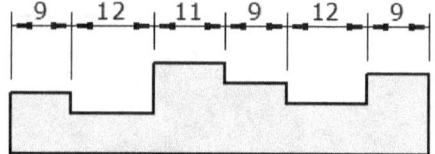

Chain dimensioning uses the end point of the previous dimension as the datum, or start point for the next dimension. It increases the chance of accumulating a tolerance error.

### 4.5.2 Datum Dimensioning:

In datum dimensioning, the measurements all originate from the same line or datum point so a tolerance error does not accumulate. More space is required on the drawing for all the dimension lines.

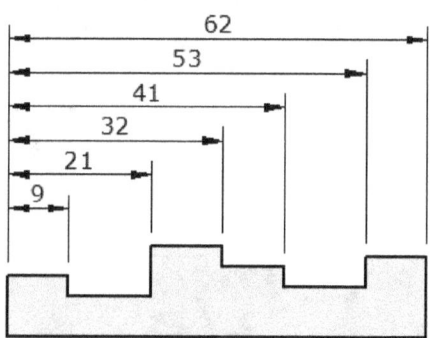

### 4.5.3 Running Dimensioning:

Datum dimensioning is similar to parallel dimensioning but takes less space. It doesn't accumulate a tolerance error either as all measurements originates from the same line. Running dimensions are not often seen in drawings as more care is needed in reading the dimensions to make sure you have the right measurement.

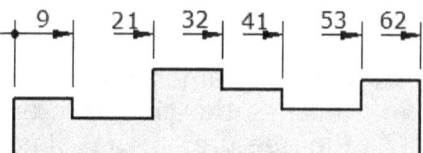

### 4.5.4 Combined Dimensioning:

Combined dimensioning is a combination of chain and parallel dimensioning; this method uses less space than parallel dimensioning and accumulates less of a tolerance error than chain dimensioning.

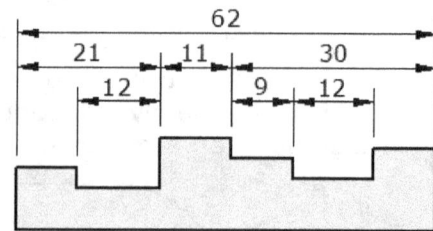

## 4.6 Dimensions "Not To Scale":

At times a dimension may be drawn out of scale due to design changes where it is less time consuming and simpler to just change the dimension, or the length of an object may be deliberately drawn shorter to view the object at a larger scale as in the case of detailing a 9 meter long steel beam.

Dimensions that are "NOT TO SCALE" are identified by a short stroke under the dimension text.

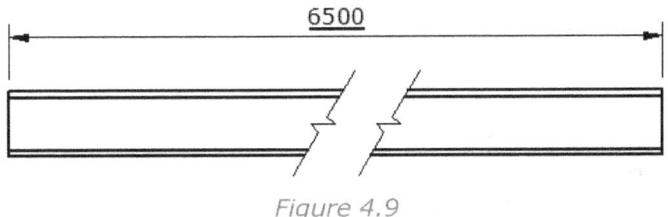

*Figure 4.9*

## 4.7 Placement of Dimensions:

Dimensions are to be placed on the drawing using the aligned or unidirectional method. In most drawing offices the aligned method is used because it can save on space and prevent dimensions running into each other as there are only two methods for placing dimensions Aligned and Unidirectional.

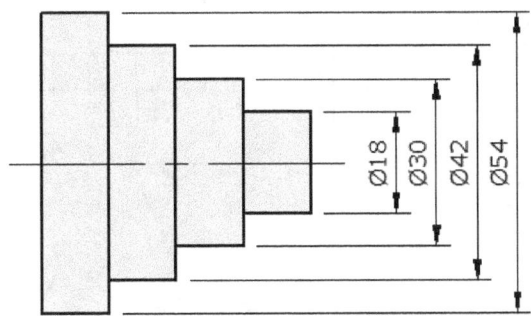

*Figure 4.10 – Aligned*

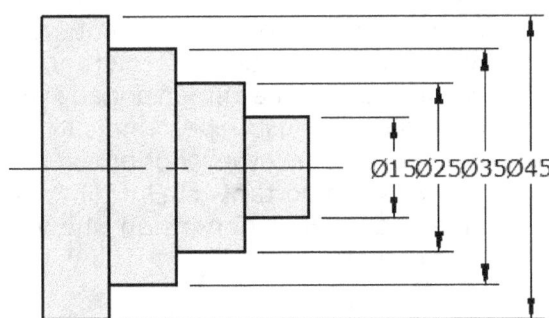

*Figure 4.11 - Unidirectional*

Figure 4.10 shows the dimensions being placed using the Aligned System while in Figure 4.11 the Unidirectional System has been used. The dimensions are unreadable in the unidirectional method because they appear to run each other forming one long number or dimension text. To overcome the problem, the dimensions can be spaced further apart however the size of the view becomes larger than required and leaving a smaller drawing area. The problem is further exasperated when the dimensions become decimals (Ø65.035).

*4.7.1 Location of Dimensions:*
When using the Aligned System of dimensioning, all horizontal dimensions are placed above and in the centre of the dimension line.

Placement on Drawing

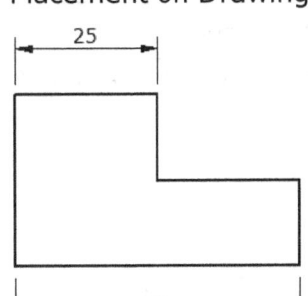

When using the Aligned System of dimensioning, all vertical dimensions are placed to the left of the dimension line so when viewed from the right they appear above the dimension line.

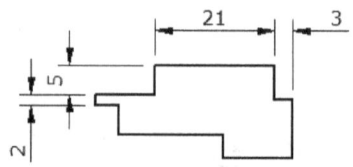

Angular dimensions must be shown with the degrees symbol in both the Aligned and Unidirectional Systems.

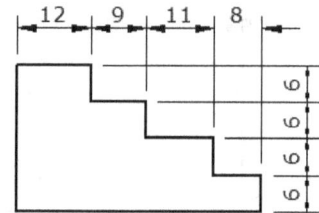

### 4.7.2 Dimensioning Small Spaces:
At times the dimension text and arrows is larger than the space; when this occurs, the arrows are placed outside the space and the text placed over one of the lines where the text cannot be confused with another dimension.

### 4.7.3 Chain Dimensioning:
Chain dimensioning is used when the function of the object will not be affected by the accumulation of tolerances. If all features are to be dimensioned, the overall dimension must be indicated within brackets. The other option is to remove the least important chain dimension and have the overall dimension shown without the brackets.

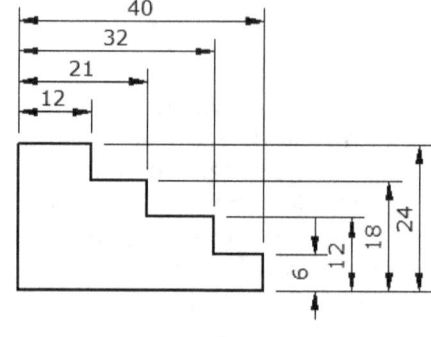

### 4.7.4 Datum Line Dimensioning:
Datum dimensioning consists of several dimensions originating from one surface. The use of datum dimensioning is more accurate than chain dimensioning because errors are not accumulated.

For large holes, either method is acceptable but most drawing offices tend to prefer the leader dimension.

The leader line MUST point to the centre of the circle.

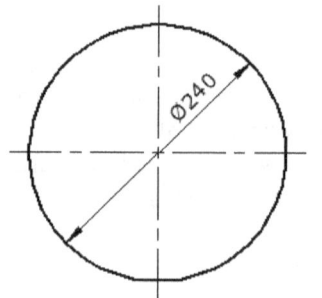

*4.7.5 Dimensioning Arcs:*
Internal leader lines must pass through, or point towards the centre of the arc.

Where the leader is applied externally, the leader line MUST point to the centre of the circle.

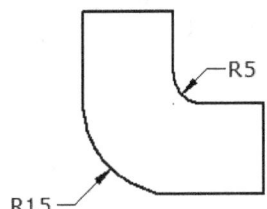

## 4.8 Dimensioning Rules:

The following list summaries briefly most of the rules which govern placing dimensions on a drawing. The rules apply to any person preparing drawings or sketches including, draftspersons, engineers, architects and tradespersons. Incorrectly placed dimension on a drawing or sketch could affect the reading and interpretation of the drawings

1. Dimensions should be kept outside the view. (Where space is limited, or the dimension is too far from the detail, it may be more efficient to place the inside the drawing).

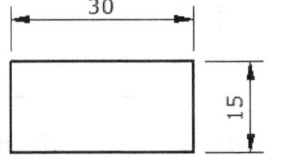  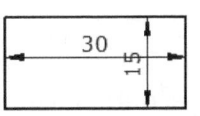 

2. The dimension for a feature is not to be repeated on a drawing. The length of an object will be shown on either the Front or Side views – not both.

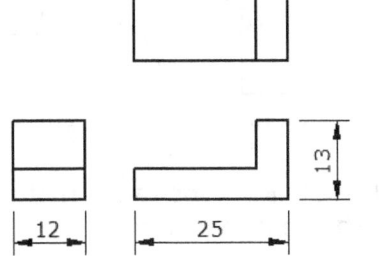  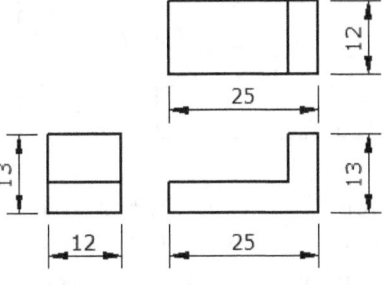 

3. Centerlines are not be used as dimension lines.

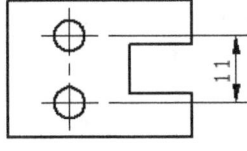  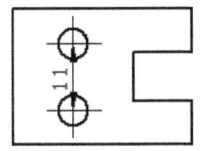 

4. Dimensions are not allowed to be applied directly to faces or lines.

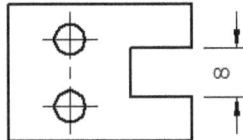  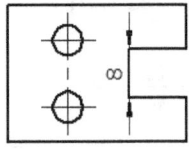

5. Any line defining the shape is not allowed to be used as a dimension line as it confuses the drawing.

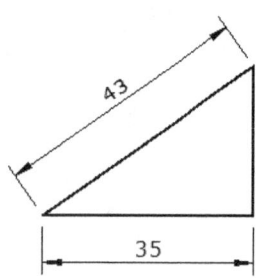  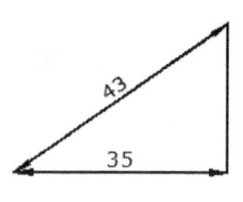 

6. Dimension lines and projection lines must not cross one another. The smallest dimension is placed nearest the outline progressing to the largest dimension on the outside.

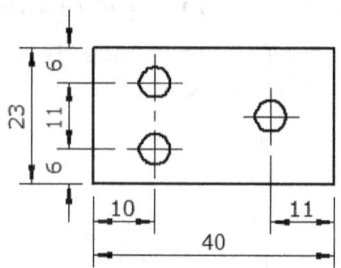 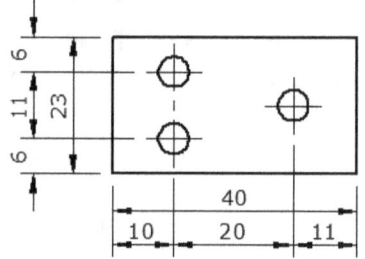 

7. Only visible outlines and centerlines are dimensioned. In extreme circumstances where a feature only appears as hidden detail, the feature may be dimensioned.

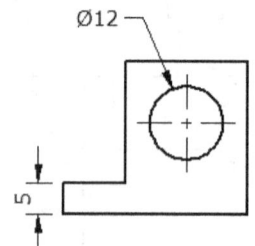 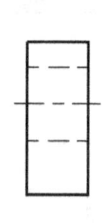  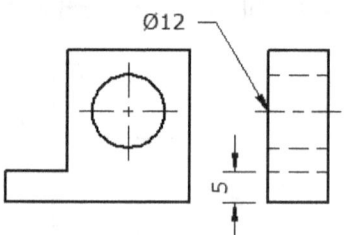 

8. When the note "DIMENSIONS ARE IN MILLIMETRES" appears on the drawing, it is not necessary to show the unit (mm) sign with the dimension.

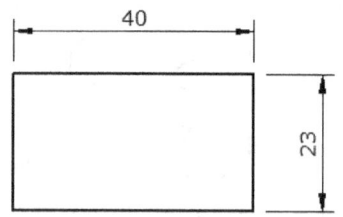 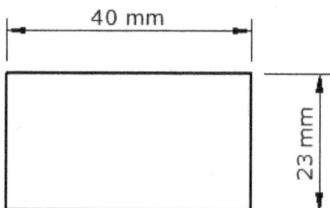 

9. Projection lines can cross one another.

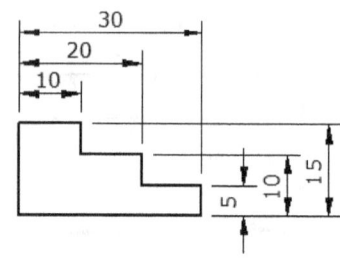  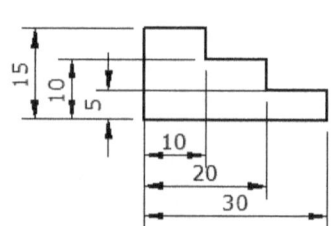

**Other rules include:**

- Dimensions should be placed in the views where the features dimensioned are shown in true shape.
- Dimensions applying to adjacent views should be placed between the views, unless clearness is better achieved by placing some of them outside.
- Do not expect the tradesperson to assume a feature is centred (as a hole in a plate); give a location dimension from at least 1 side.
- Do not expect a tradesperson to calculate or have to do any "adding up" as they may not be good at mathematics and could make an error.
- Detail dimensions should "line up" in chain dimensioning.
- Avoid a complete chain of dimensions; omit one and show the overall length.
- Centerlines (or any other line) should not cross a dimension line or the dimension text. If necessary, break the centerline each side of the dimension line and text.
- Leaders for notes should be straight and not curved, and point to the centre of circular/radial features.
- Dimension text should be centered between the arrows except when a "stack" of dimensions, the text should be staggered.
- Dimension text should never be crowded or made difficult to read.
- Notes should always be lettered horizontally on a drawing.
- Notes should be brief and clear using standard words, symbols and abbreviations.
- Holes and shafts should always be identified with the diameter (Ø) symbol before the dimension text; radii's should be identified by the letter "R" before the dimension text.
- When a dimension is not to scale, it should be underscored with a straight line.

## 4.8 Toleranced Dimensions:

Because it is virtually impossible to make a component to an exact size or finish, a tolerance is often stated which provides a range of possible close sizes. A tolerance can be described as a permissible variation in the sizes (or dimensions) of a component.

In engineering, three types of toleranced dimension are used, Unilateral, Bilateral and Direct or Limit of Size.

*Unilateral:*
A Unilateral tolerance consists of a basic size and a total allowable variation given in **ONE** direction only; the direction can be in the positive *or* negative direction, not *BOTH*.

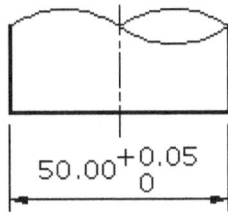
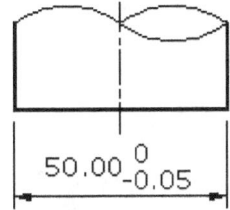

Figure 4.12    Figure 4.13

## Bilateral:

A Bilateral tolerance consists of a basic size and a small variation in size that can be either in the positive *or* negative direction as seen in .

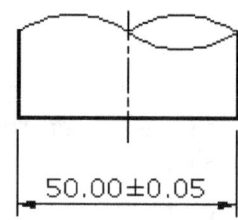

*Figure 4.14*

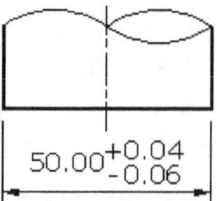

*Figure 4.15*

## Direct or Limit of size:

Limit of size specifies the absolute maximum and permissible sizes that the part can be manufactured.

Limit of Size toleranced dimensions are more commonly used of the three methods because the worker does not have to do any mental calculation, just read the measurement and determine whether the measurement lies between the maximum and minimum values.

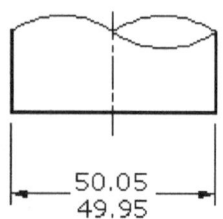

*Figure 4.16*

The upper (maximum) value is placed on top while the lower (minimum) value is placed below the dimension line.

# MEM09002B – Interpret technical drawing
## Topic 4 - Dimensions

### Skill Practice Exercises:
*Skill Practice Exercise MEM09002-RQ-0401*

1. Define a dimension.
   _____
   _____

2. Name the two types of linear system dimensioning.
   _____

3. What is the name for the type of dimensioning when all horizontal dimensions are given from the one line or feature?
   _____

4. Name the features of a dimension.
   _____

5. What is incorrect with the dimensions shown in the image below?

   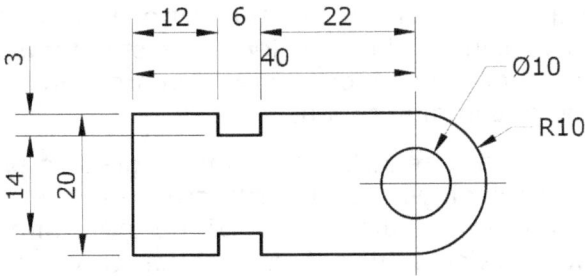

   _____

6. What are the normal sizes of a dimension arrow?
   _____

7. What does a dimension mean if there is a short line directly below the dimension text?
   _____

8. Name six basic factors in dimensioning.
   _____
   _____

# Topic 5 – Orthographic Projection:

### Required Skills:
- Name the methods used to produce engineering drawings.
- Identify projection symbols and distinguish the difference between First Angle and Third Angle Projection.
- Minimum number of views required to describe the component/assembly.

### Required Knowledge:
- Application of AS1100.
- Relationship between the views contained in the drawing.
- Objects represented in the drawing.

**NOTE:**
This topic is intended to introduce the student to the types of projection, names and positions of the views. The drawing of orthogonal views in engineering drawings is covered in units MEM09003B - Prepare basic engineering drawing MEM09005B - Perform Basic Engineering Drafting, and other drafting discipline related units.

## 5.1 Orthographic Projection:
Most drawings produced and used in industry are multiview drawings. Multiview drawings are used to provide accurate three-dimensional object information on two dimensional media, a means of communicating all of the information necessary to transform an idea or concept into reality. The standards and conventions of multiview drawings have been developed over many years, which equip us with a universally understood method of communication.

Multiview drawings usually require several orthographic projections to define the shape of a three-dimensional object. Each orthographic view is a two-dimensional drawing showing only two of the three dimensions of the three-dimensional object. Consequently, no individual view contains sufficient information to completely define the shape of the three-dimensional object. All orthographic views must be looked at together to comprehend the shape of the three-dimensional object. The arrangement and relationship between the views are therefore very important in multiview drawings.

Orthographic projection is a system of drawing to represent 3-D objects by using multiple view drawings. The word "Ortho" is a Greek word that means right or true. In this system of projection, the 3-D object is projected perpendicularly onto a projection plane with parallel projectors as shown in Figure 5.1.

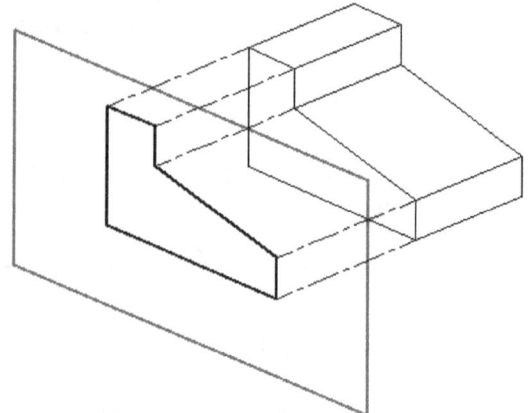

*Figure 5.1*

## Topic 5 - Orthogonal Projection

### 5.1.1 Basic Views:

All objects can be projected in six orthogonal directions (Figure 5.2). The resulting views are called basic views. Orthogonal Projection can be thought of as a 3D object being placed inside a transparent box, and views projected orthogonally onto the six walls of the box.

The basic views are:

- Front View
- Top View
- Right & Left Views
- Bottom View
- Rear View

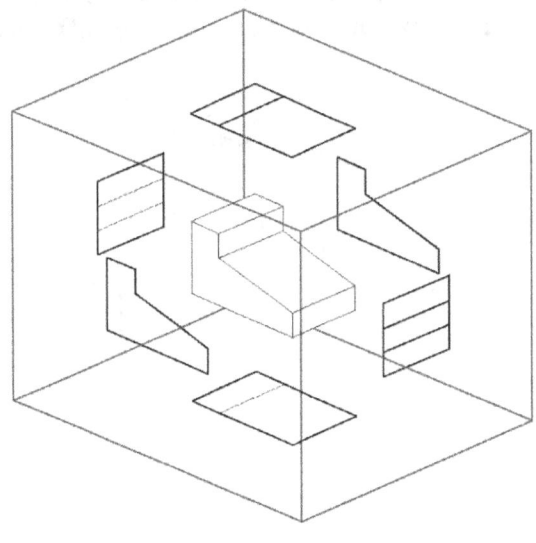

*Figure 5.2*

### 5.1.2 Developing the Box:

The transparent box may suit a "virtual Reality" environment; it cannot be placed on a drawing or forwarded to the workshop. To make sense on the drawing, the box is opened or spread-out onto a common plane which is the drawing sheet.

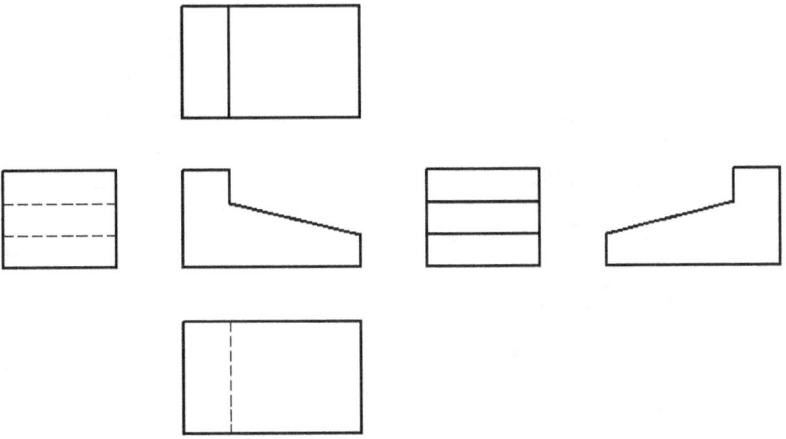

*Figure 5.3*

Figure 5.3 shows how the drawing would look like after cutting and spreading out the box

### 5.2 Projection Systems:

Two methods of projection have been used to produce engineering drawings; Third Angle Projection is the preferred method stated in AS 1100 and is used throughout Australia in most drafting disciplines; however, some drafting disciplines still tend to use First Angle Projection. The difference between Third and First Angle Projection is the position of the Side, Top and Bottom Views in relation to the Front View. Until around 1890 all countries produced drawings in First Angle Projection, modern multi-national offices work entirely in Third Angle Projection.

### 5.2.1 Third Angle Projection:

The plane of projection lies between the observer and the object.

When the views are drawn, the Top View is located ABOVE the Front View, the Left Side View is located to the LEFT of the Front View, the Right View is located to the RIGHT of the Front View, and the Bottom View is located directly BELOW the Front View.

**N.B. The views are drawn from where the object is being viewed. Viewed from the left and drawn on the left; viewed from on top and drawn on top.**

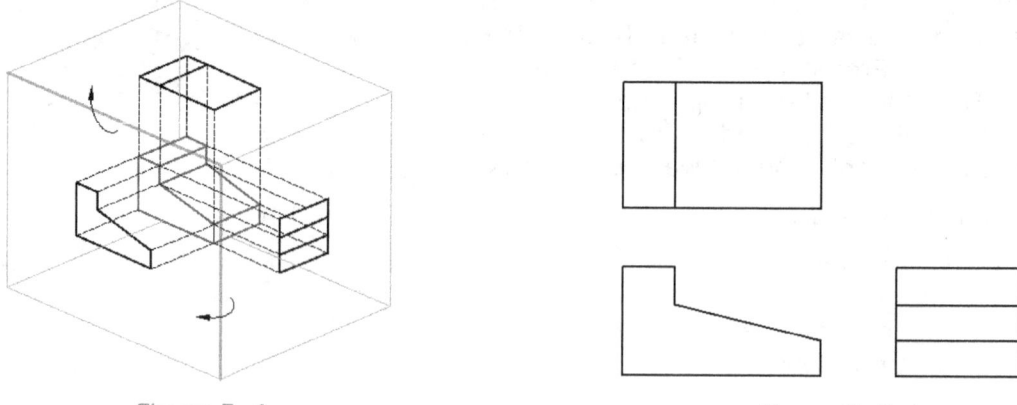

Figure 5. 4                                   Figure 5. 5

### 5.2.2 First Angle Projection:

The object lies between the observer and the plane.

When the views are drawn, the Top View is located BELOW the Front View, the Left Side View is located to the RIGHT of the Front View, the Right View is located to the LEFT of the Front View, and the Bottom View is located directly ABOVE the Front View. The location is exactly opposite to Third Angle Projection.

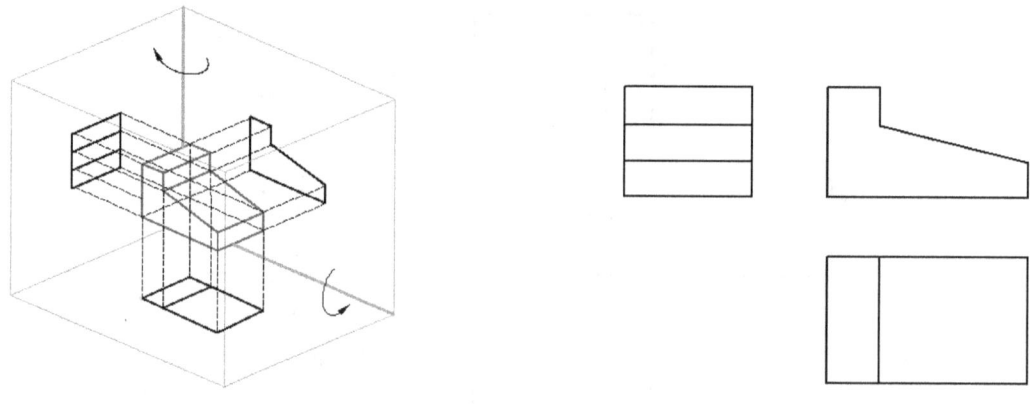

Figure 5. 6                                   Figure 5. 7

### 5.2.3 Projection Symbols:

To avoid misunderstanding, international projection symbols, as shown in Figure 5.8 and Figure 5.9, have been developed to distinguish between Third and First Angle projections on drawings.

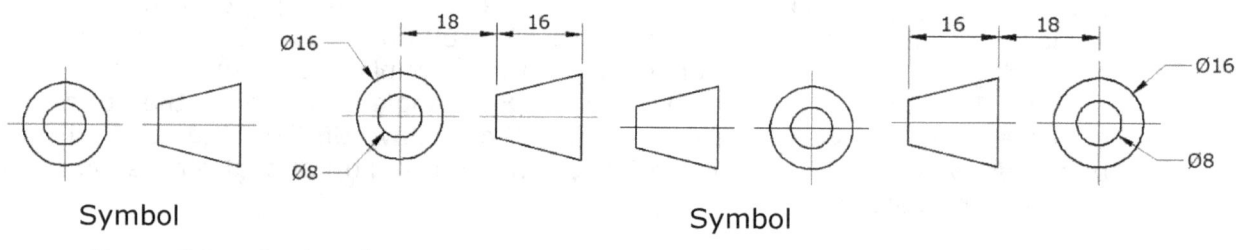

Symbol                                        Symbol

Figure 5.8 – Third Angle Projection           Figure 5.9 – First Angle Projection

## 5.3 Number of Views:

The number of views required depends on the complexity of the component; some drawings may require only one view with the width of the material shown under the Title while other components may require 5 or 6-views to fully describe the object. Figure 5.10 shows a complex cam that requires only 1-view to fully show all the features and dimensions; the thickness is constant and Side View would only show a rectangle so the thickness can be placed below the Title.

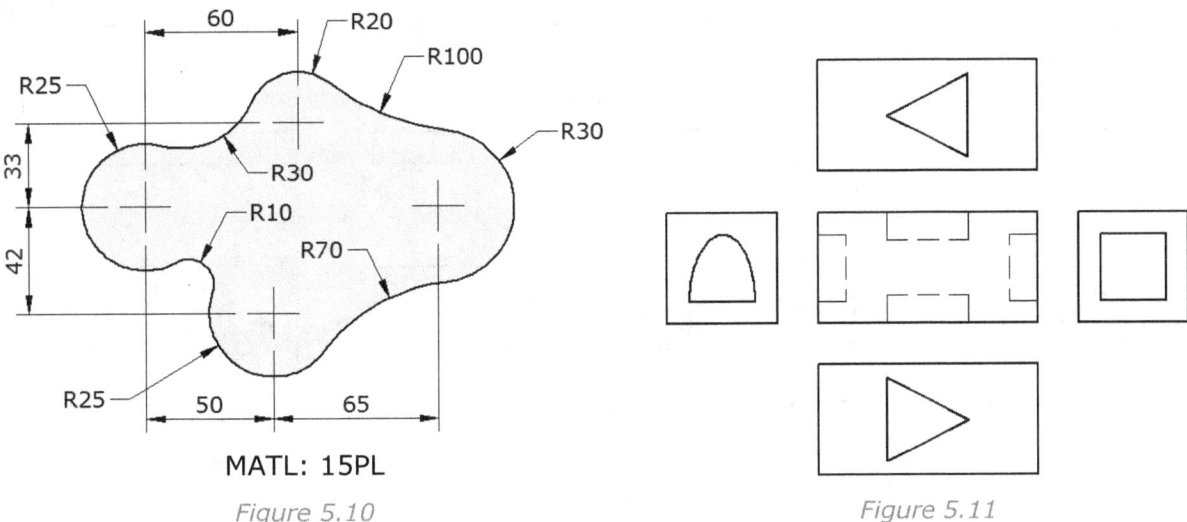

Figure 5.10

Figure 5.11

Figure 5.11 shows a simple hypothetical block with a series of different shape holes; however 5-views are required to fully describe the shape.

Most detail drawings require 3-views to fully display the shape of the component and its dimensions however some components could only require 2-views, especially if it is symmetrical.

**N.B. The number of views depends on the complexity of the object being drawn.**

# MEM09002B – Interpret technical drawing
## Topic 5 - Orthogonal Projection

### Skill Practice Exercises:

*Skill Practice Exercise MEM09002-SP-0501*
Identify the correct projection angle, or no projection used in the following examples.

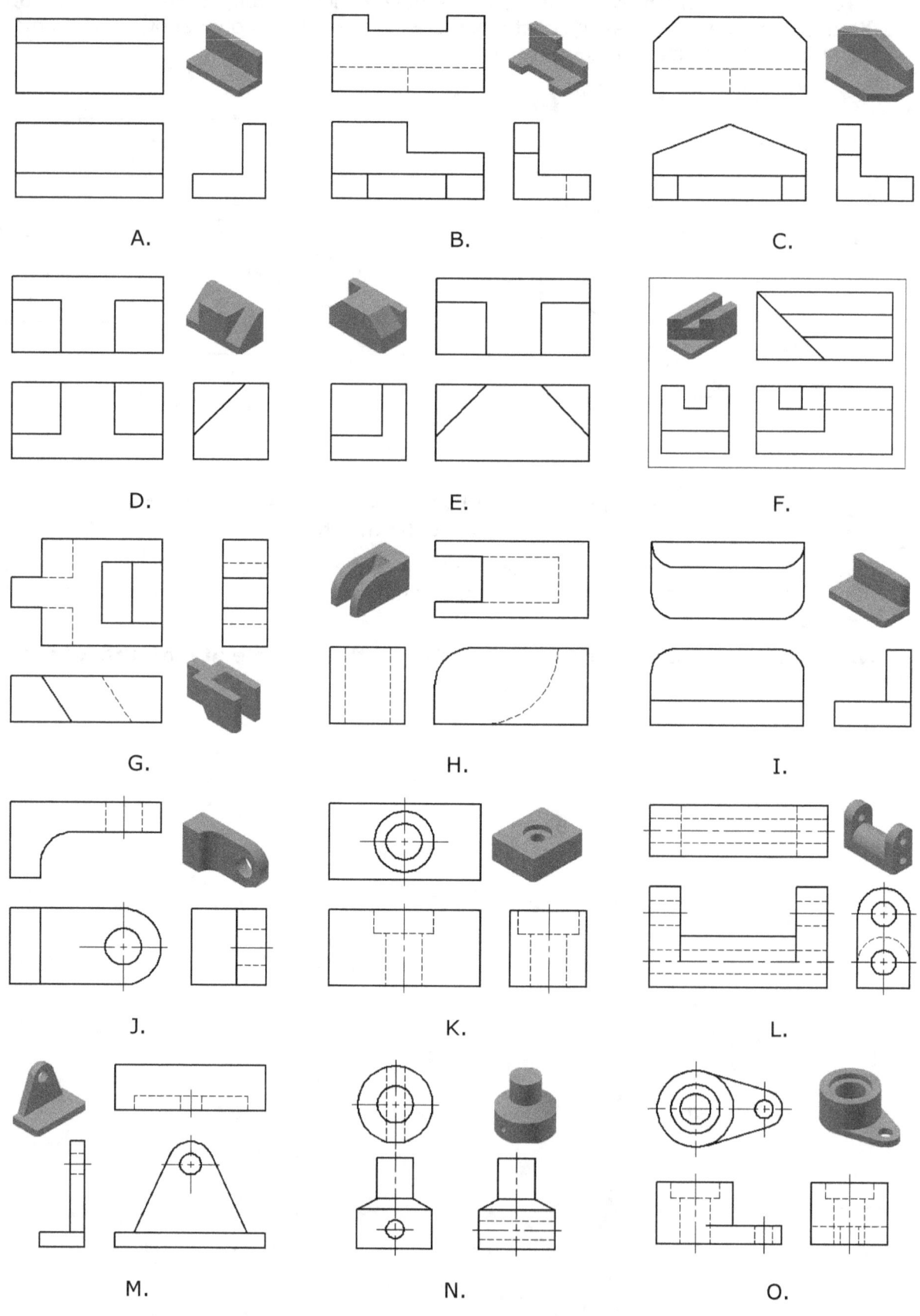

MEM09002B – Interpret technical drawing

Topic 5 - Orthogonal Projection

**MEM09002-SP-501 Answer Sheet:**

A. _____

B. _____

C. _____

D. _____

E. _____

F. _____

G. _____

H. _____

I. _____

J. _____

K. _____

L. _____

M. _____

N. _____

O. _____

Name: _____

# MEM09002B – Interpret technical drawing
## Topic 5 - Orthogonal Projection

*Skill Practice Exercise MEM09002-SP-0502*

Identify the correct projection angle used in the following examples. Answers are First Angle Projection, Third Angle Projection, Incorrect (First & Third Combined).

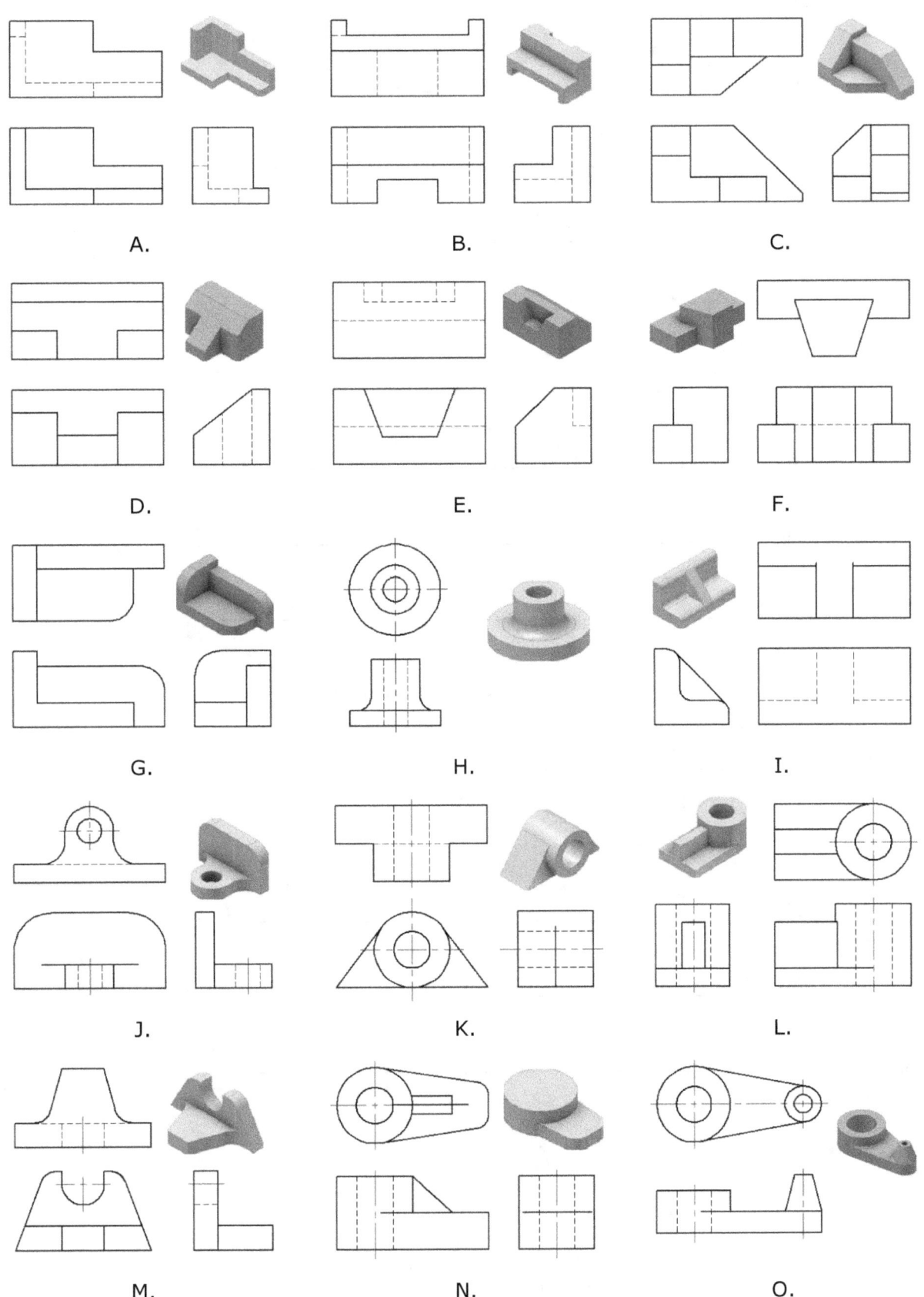

A.  B.  C.

D.  E.  F.

G.  H.  I.

J.  K.  L.

M.  N.  O.

MEM09002B – Interpret technical drawing

## Topic 5 - Orthogonal Projection

**MEM09002-SP-502 Answer Sheet:**

A. _____

B. _____

C. _____

D. _____

E. _____

F. _____

G. _____

H. _____

I. _____

J. _____

K. _____

L. _____

M. _____

N. _____

O. _____

Name: _____

# Topic 6 – Sections:

## Required Skills:
- Name the various types of sections appearing on a drawing.
- Identify different components and materials on an assembly drawing.
- Associate cutting planes with the appropriate sectional view on the drawing.

## Required Knowledge:
- Methods of projection.
- Relationship between the views contained in the drawing.
- Objects represented in the drawing.
- Understanding of webs, ribs and thin components.

## 6.1 Introduction:

In an orthographic projection drawing, outlines and edges of an object are usually depicted with continuous lines and internal details are normally illustrated by using hidden lines. When dealing with complex objects, there may be many hidden lines and these hidden lines may become very confusing so sectioning techniques are used to cut segments out of the object to clearly show internal details.

Using sectioning technique, the object is imagined to be cut by a plane and the portion nearer to the observer is imagined to be removed; this way, the interior is exposed and is shown in continuous lines. To specify the cutting, hatching is applied at the plane of cutting where any material would be cut.

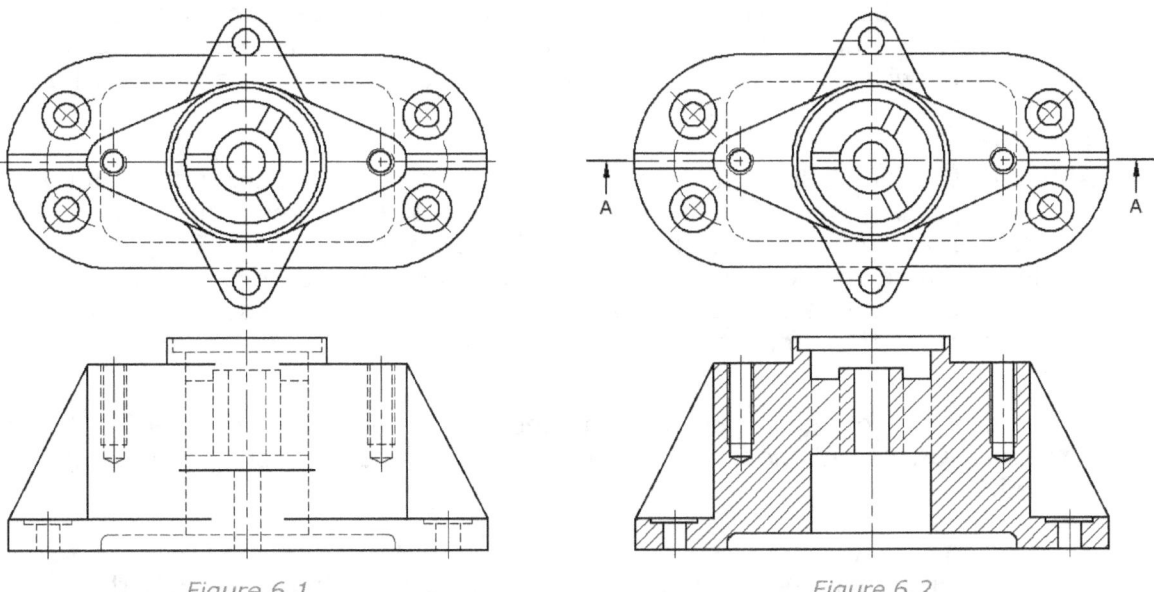

Figure 6.1                    Figure 6.2

Figure 6.1 shows the Front & Top Views of a Valve Body Cover; the visible outlines fully describe the shape of the component however the internal detail can become confusing, especially for a person with limited knowledge of reading technical drawings. The same component is shown in Figure 6.2 however the component has been cut along the longitudinal centreline and the front part removed; the Front View has been replaced by a Full Sectional View taken along the centreline. Although the triangular webs are cut in theory, they are not hatched, however, the internal webs holding the collar have been hatched using an alternate method of sectioning to improve the understanding of the detail.

## 6.2 Cutting Plane:

The cutting plane indicates where the section is being taken and is normally indicated by a centreline with short heavy lines at each end, with an arrow pointing in the viewing direction and a distinguishing letter or combination letter and number.

Figure 6.3                    Figure 6.4

Modern drawings use the cutting plane shown in Figure 6.3 while older drawings may indicate the cutting plane as shown in Figure 6.4. The arrows on the cutting plane MUST lie in the correct projection angle used on the drawing.

NOTE: A Sectional View should be located where possible; however, it is not compulsory especially if an end view is already used. The sectional view MAY be placed anywhere on the drawing, even on another sheet attached to the drawing number.

While the normal orthogonal view may not be labelled, Sectional Views must be labelled. The letter/number identifying the view can consist of a letter commencing with "A", "B", "C" etc. or consist of a letter and a number, "3-E". The combination of a letter and a number could refer to the grid zone reference of the view e.g. Zone 3-E, or the number could refer to a sheet in the drawing set followed by the identifying label e.g. Sheet 3, Sectional View E.

## 6.3 Cross Hatching:

Any material which would actually be cut if the component was physically cut along the cutting line is indicated by thin continuous lines, which are equally spaced, and at the same angle. The cross hatching on sectional views detailing a single component normally have the hatching spaced at approximately 3 mm and at an angle of 45° as shown in Figure 6.2. In assembly drawings where there are multiple components, the spacing of the hatching will vary with the spacing of larger components at a large space (possibly 5 mm) and the small spaces for the smaller components (1 mm); the angle of the hatching will also alternate from the forward angle (45°) to a backward angle (135°).

## 6.4 Types of Section:

Although the different sections and sectional views have been named for identification (e.g. Section A-A) and for specifying the type of the view required in a drawing, the names are not shown on the drawing for the same reason that a Top, Front, or Side View would not be so labelled. The views are easily interpreted and a competent worker does not require the name to read the drawing. The names are assigned by the character of the section or the amount of view in the section, not by the amount of the object removed.

The main types of section used in engineering drawings are Full, Half and Part Sections however, Offset, Aligned, Revolved, Removed and Broken Sections are commonly utilised.

### 6.4.1 Full Section:

A full section view is made by passing an imaginary cutting plane fully through an object. The figure shows an imaginary cutting plane passing fully through an object and the front half removed. The view section is not necessarily restricted to the Front View but could also be the Top or Side views; depending on the complexity of the component or assembly, two or more sectional views may be required.

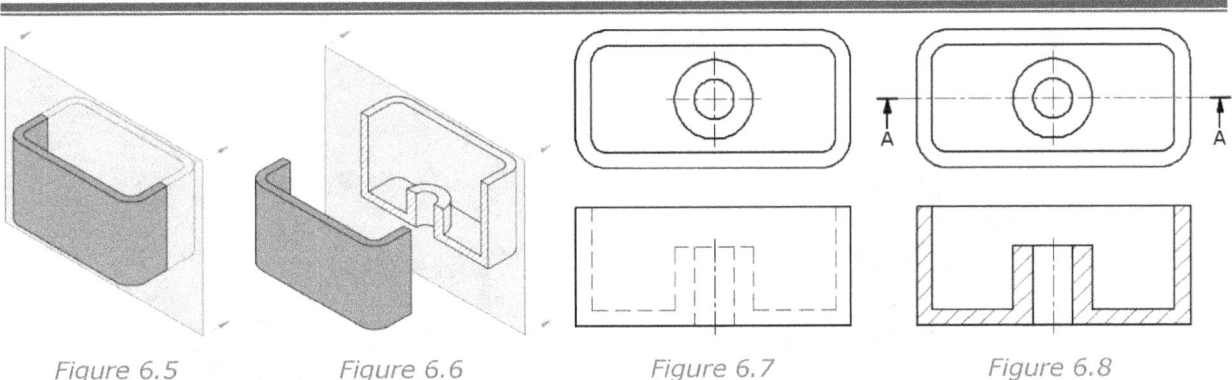

*Figure 6.5*    *Figure 6.6*    *Figure 6.7*    *Figure 6.8*

Figure 6.5 shows a Casing which has the cutting plane located along the longitudinal centreline. In Figure 6.6, the front part of the casing has been moved backwards after the "imaginary" cut has taken place along the centreline.

Figure 6.7 displays the standard Front and Top Views when drawn as an orthogonal projection with the internal detail indicated by hidden (dashed) outlines. Figure 6.8 shows the front piece removed to display the collar in full detail; the parallel lines are placed ONLY where material would be cut by the "imaginary" saw.

**6.4.2 Half Section:**
Half sections are created by passing an imaginary cutting plane halfway through an object and one quarter of it is removed. Hidden lines are omitted on both halves of the section view. External features of the part are drawn on the unsectioned half of the view. A centre line, not an object line, is used to separate the sectioned half from the unsectioned half of the view. Half section views are most often used on parts that are symmetrical, such as Pulleys and Couplings etc.

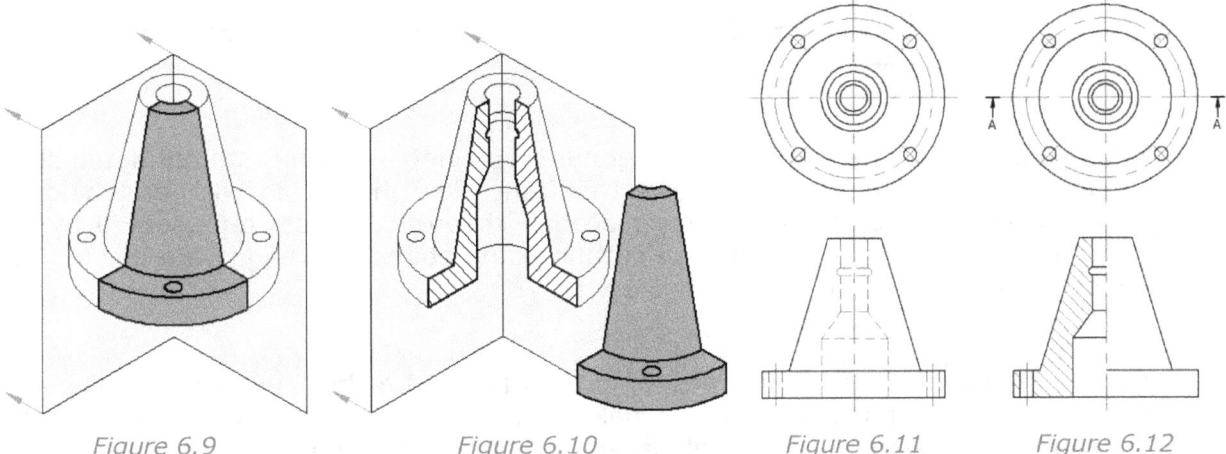

*Figure 6.9*    *Figure 6.10*    *Figure 6.11*    *Figure 6.12*

Figure 6.9 shows a Tapered Base with a cutting plane passing along the centrelines while in Figure 6.10 a quarter of the component has been removed to show the internal detail.

Figure 6.11 shows the standard view where the internal detail is represented by hidden outlines. As the Tapered Base is symmetrical, a Half Sectional view can be used to describe the detail as shown in Figure 6.12. Note there is no hidden detail shown in the sectioned view and the cutting line is represented as a centreline, and NOT a visible outline, because the cut is only imaginary.

**6.4.3 Offset Section:**
An offset section has its cutting plane bent at 90° angles to pass through important features. Offset sections are used for complex parts that have a number of important features that cannot be sectioned using a straight cutting plane. The cutting plane is bent at 90° to pass through the hole, then bent 90° again to pass through the slot.

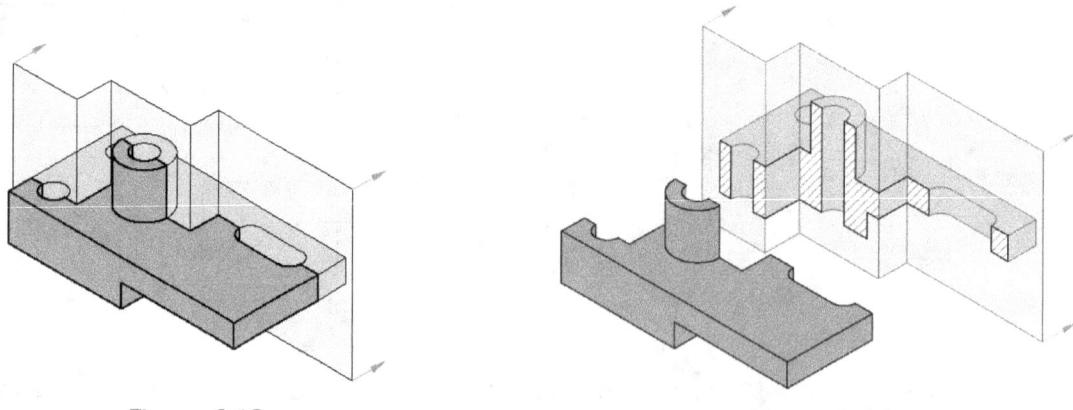

Figure 6.13                                              Figure 6.14

Figure 6.13 shows a Slotted Drill Jig with a cutting plane passing through the centre of the circular holes in the base and collar, and the slotted hole in the base. In Figure 6.14, the component has been cut along the cutting plane and the front removed to clearly show the internal details.

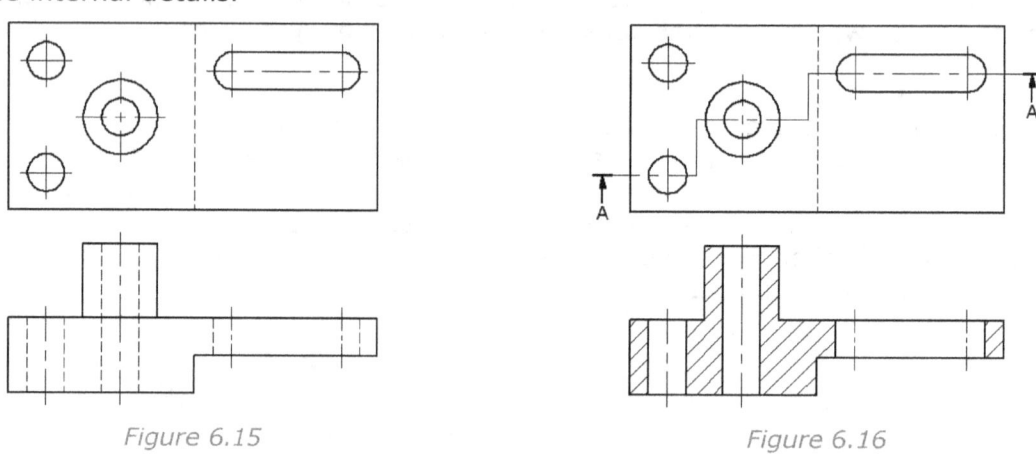

Figure 6.15                                              Figure 6.16

Figure 6.15 shows the standard orthogonal Front View with the internal detail shown as hidden outlines. Figure 6.16 shows the front view in section with and hidden lines replaced by visible outlines and any material "theoretically cut" represented by a hatch pattern. Note the lines forming the cutting plane at the 90° bends are not indicated in the sectional view as the cut is imaginary only.

### 6.4.4 Aligned Section:

Aligned sections are special types of orthographic drawings used to revolve or align special features of parts to clarify or make them easier to represent in section. Normally the alignment is along a horizontal or vertical centreline and always less than 90°.

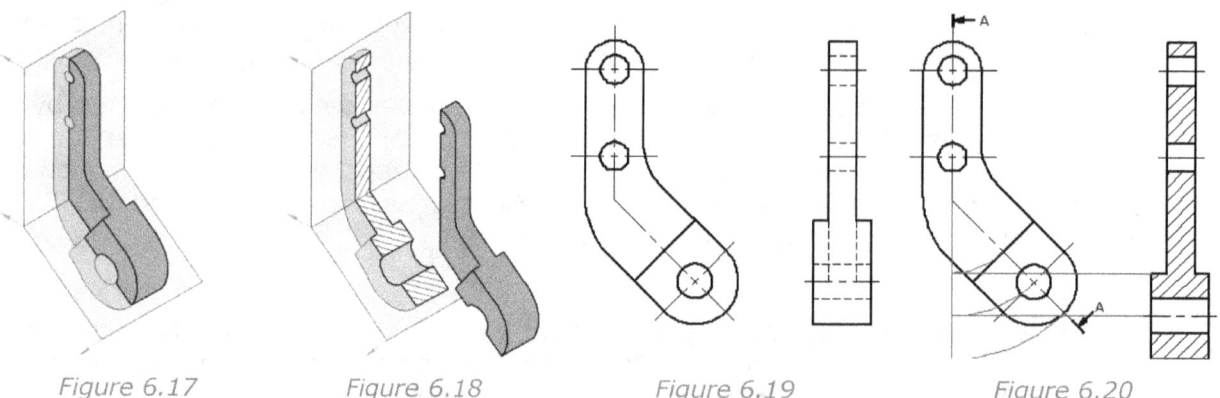

Figure 6.17      Figure 6.18      Figure 6.19      Figure 6.20

Figure 6.17 shows a Lever Arm where the cutting plane passes along the centreline. Figure 6.18 shows the area in front of the cutting plane removed to view the internal detail. Figure 6.19 is an orthogonal view of the arm where the visible and hidden details are projected across to the side view. Figure 6.20 however shows the height of the sectional view the true distance along the cutting plane. The location of the features is determined by rotating the feature to an "imaginary" vertical centreline and then projecting across to the sectional view.

### 6.4.5 Revolved Section:
A revolved section is made by revolving the cross-section view of the part about an axis of revolution and placing the section view on the part. The cross section created at the position that the cutting plane passed is then revolved 90° and drawn on the view. Visible lines adjacent to the revolved view can either be drawn or broken out using conventional breaks.

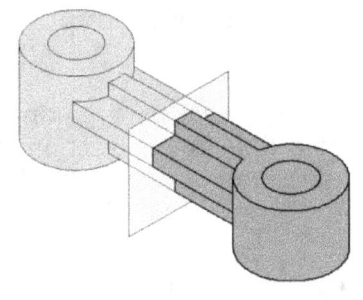

Figure 6.21

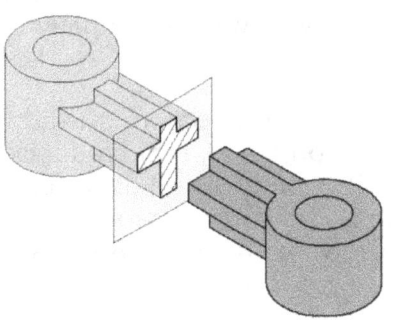

Figure 6.22

Figure 6.21 shows a Link with a stiffener forming a cross joining the two ends with a cutting plane passing through the centre of the stiffener. In Figure 6.22, the area in front of the cutting plane has been removed to view the stiffener.

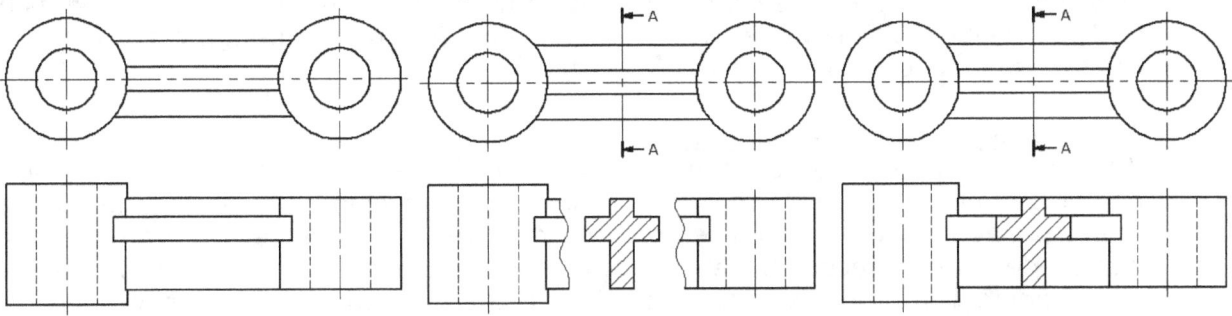

Figure 6.23　　　　　　　Figure 6.24　　　　　　　Figure 6.25

Figure 6.23 shows the orthogonal views of the Link, however, the exact shape or details of the stiffener are indistinct. Figure 6.24 and Figure 6.25 show that the sections have been revolver on their axes. In Figure 6.24 break lines have been included with the lines between deleted showing the detail of the stiffener with no lines confusing the detail. Figure 6.25 shows the same detail of the stiffener but the break lines have been omitted. Either method is allowable on a drawing.

### 6.4.6 Removed Section:
Removed sections are used to show the contours of complicated shapes such as wing and fuselage, blades for jet engines. Removed sections are made in a manner similar to revolved sections, by passing an imaginary cutting plane perpendicular to a part then revolving the cross section 90°; however, the cross section is then drawn adjacent to the orthographic view, not on it. If a number of removed sections are done on a part, cutting plane lines may be drawn with labels to clarify the position from which each section is taken. Whenever possible, a removed section should be on the same sheet as the part it represents, and it should be clearly labelled.

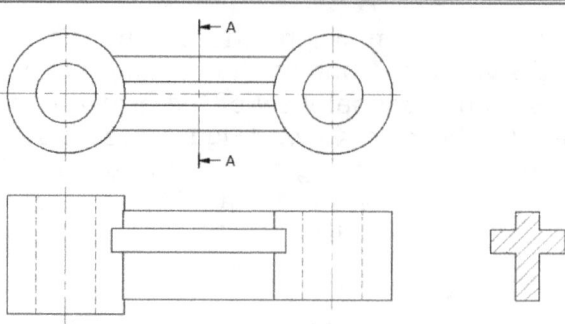

*Figure 6.26*

The Link shown in Figure 6.21 is used again to demonstrate a Removed Section; instead of the section being located in the sectional view, the section is located horizontally in line as shown in Figure 6.26. If several sections along a feature are to be removed, the sectional views are normally located directly above and in line with the cutting plane.

### 6.4.8 Broken Section:

A broken-out section is used to show interior features of a part by breaking away some of the object. A broken-out section is used instead of a half or full section view to save time. A break line separates the sectioned from un-sectioned half of the view. A break line is drawn free-hand to represent the jagged edge of the break. No cutting plane line is drawn with a broken-out section view.

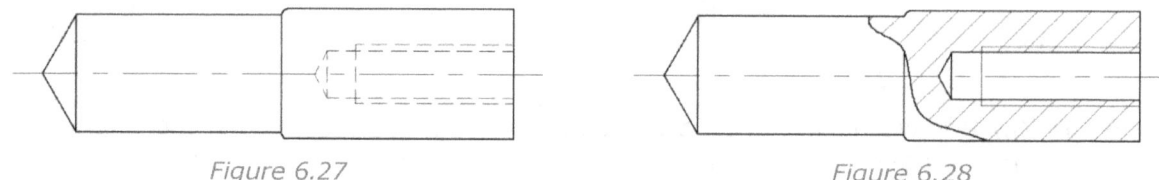

*Figure 6.27*          *Figure 6.28*

Figure 6.27 shows a simple Plumb Bob with a tapped hole in one end displayed as hidden outlines; Figure 6.28 however has a short break line drawn through the throat to detail the tapped hole for dimensioning.

## 6.5 Webs, Ribs & Thin Sections:

When a cutting plane passes longitudinally through the centre of a web or rib, the cross-hatching is eliminated from the webs as if the cutting plane was passing just in front of them, or, as if they had been temporarily removed and then replaced after the section was made. A true sectional view with the webs cross-hatched gives a misunderstanding to the perceived thickness of the web. If the cutting plane passes transversally, that is, at right angles to the length or axis direction, a hatch pattern is ALWAYS used.

The Post Support shown in Figure 6.29 has the hatching applied in accordance with standard and conventional techniques used in the majority of Drafting Offices; the hatching gives the impression of density or thickness where the hatching is located while the triangular webs appear thinner.

Figure 6.30 on the other hand gives the impression the shape is more conical with no webs.

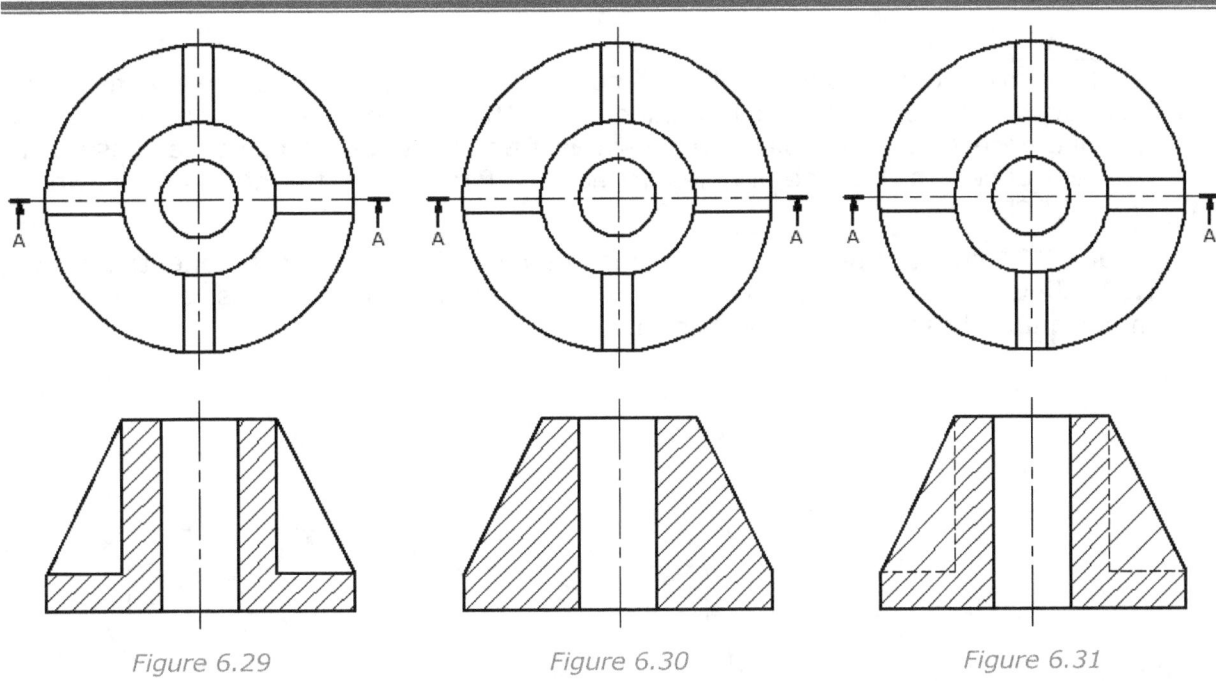

Figure 6.29    Figure 6.30    Figure 6.31

**6.5.1 Alternate Method:**
In some cases, omitting the cross-hatching of the webs or ribs can provide inadequate detail or an ambiguous illusion as shown in Figure 6.32.

Figure 6.31 shows the alternate method to indicating the webs or thin sections. The inner lines where the web joins with the main body of the component are displayed as hidden lines and while the main body of the component is hatched using a normal spacing of the hatching, the hatching on the web is drawn at twice the scale with every second hatching line extending through the web.

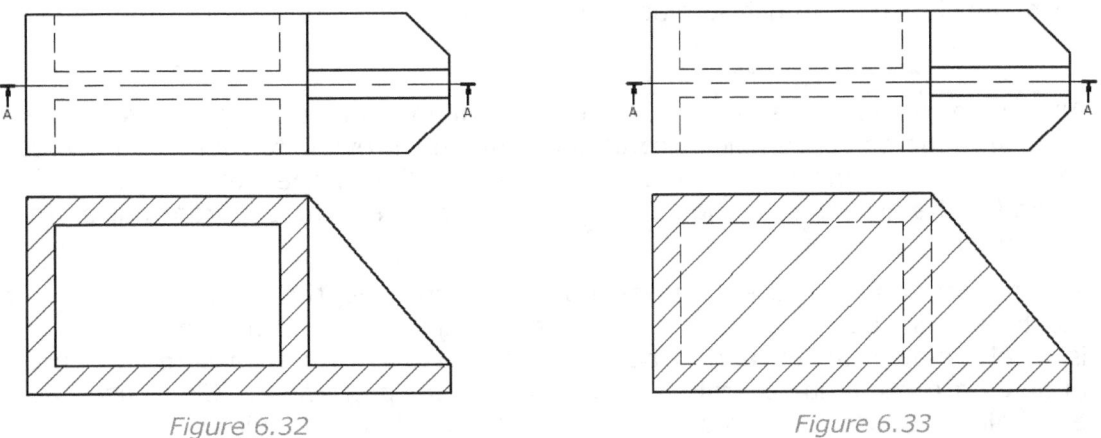

Figure 6.32    Figure 6.33

Compare the two images above, both are CORRECT, however, Figure 6.32 appears to be hollow with no internal web; the web can only be identified as existing in the top view and if there were more hidden detail, the web may be missed. In Figure 6.33, it is apparent at first glance a web is included in the component.

## 6.6 Holes:

As there are often an odd number of holes used to locate or secure a component, and hole located in front, or directly behind the cutting plane MAY be included in the view if deemed necessary. The holes are shown as if an offset cutting plane was passed through the centrelines; in the case of holes drilled on a Pitch Circle, the holes are shown on their correct PCD.

Figure 6.34 shows a section taken along the centreline of the component; the holes are NOT shown. The drafter in creating Figure 6.35 however has deemed the holes important and has included them in the view.

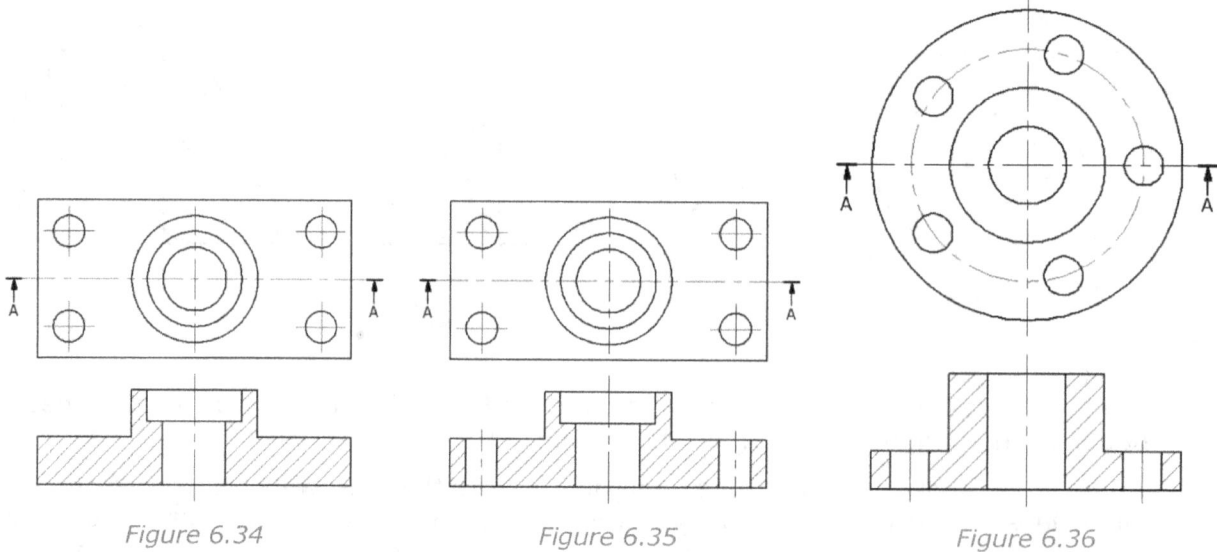

Figure 6.34          Figure 6.35          Figure 6.36

Figure 6.36 shows a base with five holes equally spaced on a designated pitch circle diameter (PCD). A only one hole is located on the cutting plane, the second hole is normally shown but must be placed on the true PCD.

## 6.7 Sectioned Assembly Views:

An assembly section is required when a combination of components are assembled in their working (or assembled) condition. All the previously mentioned types of section can be used to improve the clarity and readability of the detail drawing. The cutting plane for the assembly section is often offset to reveal the separate parts of the machine or structure.

The purpose of the assembled section is to improve the clarity of the interior of the machine or structure so that the parts can be clearly shown and identified. Hidden detail is rarely shown, an exception being a tapered pin through assembled parts; if hidden outlines are required, a broken section would be preferable. Clearances between bolts and holes are rarely shown apart from small scaled detailed views.

The cross-hatch MUST be different for each component and is achieved by adjusting the scale and angle of the pattern as shown in Figure 6.38, Standard symbols shown in Figure 6.37 are used in most engineering offices however, if a symbol for material is different to that shown or not included in the standards, then the symbol must be included in Key List on the drawing.

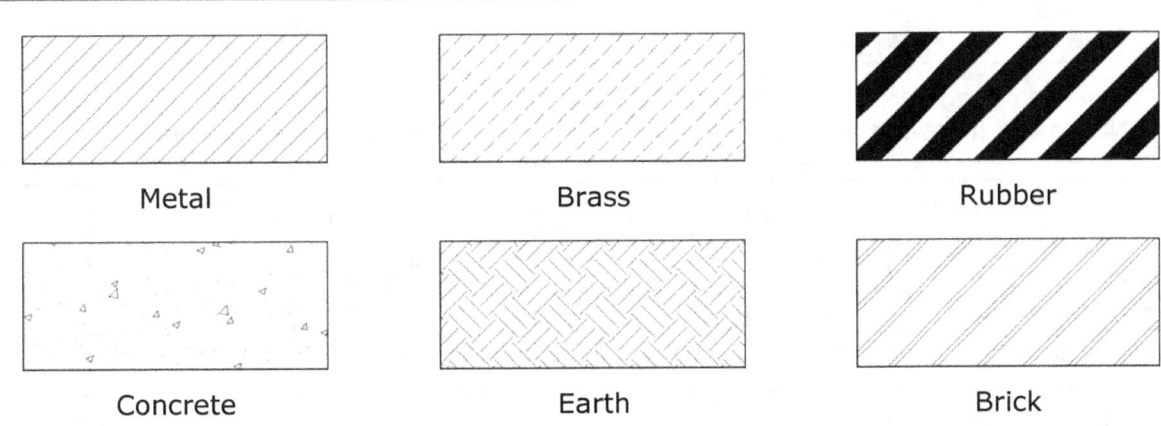

Figure 6.37

Although hatching must be at different angles and scales for different components, it must be identical when the same component is featured elsewhere on the drawing and view; the bush in Figure 6.38 is a typical example.

## 6.8 Parts Not in Section:
The reason for sectional view is to clearly identify internal detail. Many machine elements such as fasteners, pins, and shafts, have no internal detail and therefore do not require sectioning. Although the parts may appear inside an assembled view, the parts are NOT hatched because there is no detail to show; an exception to the rule is when a hole or keyway is required in the component and then a broken section would suffice.

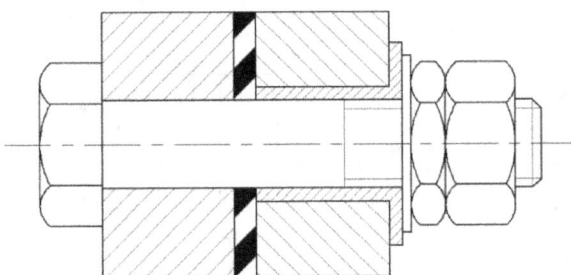

Figure 6.38

Figure 6.38 shows a simple assembly where two plates of dissimilar metals are separated by an insulating strip and bush, and then fastened using bolts, washers, nuts and locking nuts. The fasteners are not sectioned and the hidden outlines of bolt are omitted in way of the nuts and washer. All other components are hatched.

# MEM09002B – Interpret technical drawing
## Topic 6 - Sections

### Skill Practice Exercises:

*Skill Practice Exercise MEM09002-SP-0601:*
Name the following various types of sectional views in the space provided.

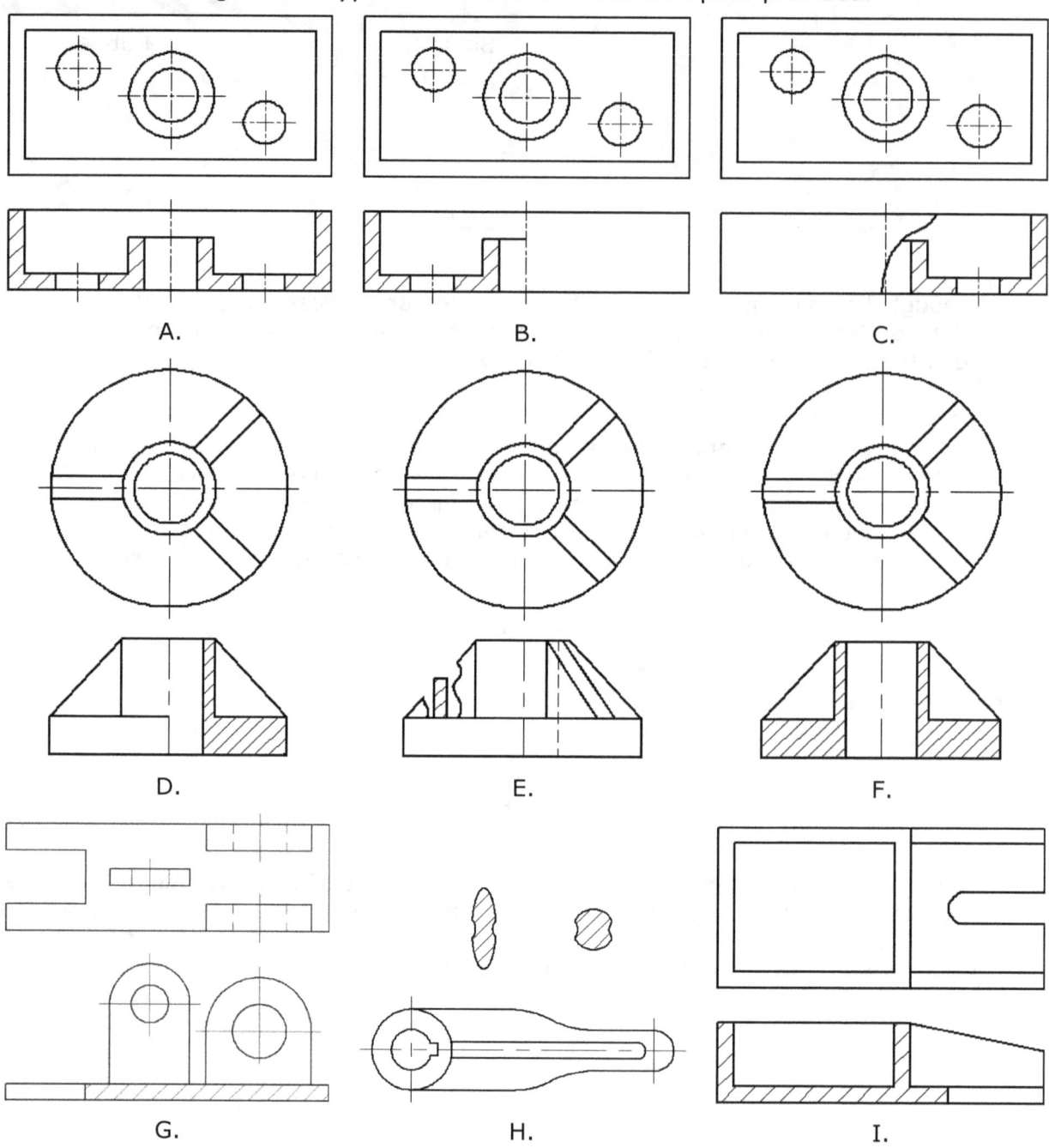

# MEM09002B – Interpret technical drawing
## Topic 6 - Sections

**MEM09002-SP-0601 Answer Sheet:**

A. _____

B. _____

C. _____

D. _____

E. _____

F. _____

G. _____

H. _____

I. _____

Name: _____

MEM09002B – Interpret technical drawing

Topic 6 - Sections

*Skill Practice Exercise MEM09002-SP-0602:*
Select the correct sectional view from the following:

## MEM09002-SP-0602 Answer Sheet:

1. _____

2. _____

3. _____

4. _____

5. _____

Name: _____

# Topic 7 – Scales:

## Required Skills:
- Interpret views of varying scales on a drawing.
- Measure an object using a scale rule.
- Undertake numerical operations to convert scaled sizes to full size measurements.

## Required Knowledge:
- Standard scales used in engineering drawing.
- Units of measurement used in the preparation of the drawing.
- Dimensions of the key features of the objects depicted in the drawing.

 **NOTE: NO DIMENSION SHOULD EVER BE LIFTED DIRECTLY FROM A DRAWING. DO NOT SCALE FROM A PRINTED DRAWING.**

## 7.1 Introduction to Scales:
Draftspersons produce drawings to fit on standard size sheets (A0, A1, A2, A3, & A4). In the days prior to computers, a draftsperson had to spend time before any drawing, planning the number and location of views, and the size required to fully describe the view. Every view was required to be drawn at a certain scale, either drawing the view at full size, or reducing the, or enlarging the view.

The scales used in an office is determined mainly by the drafting discipline, mechanical use scales from 2:1 to 1:20, instrumentation drafting may use 50:1 to 1:1, while architectural use 1:2 to 1:500 and topographical and hydrology offices may use 1:2,000 to 1:1,000,000.

In modern Drawing Offices, the use of CAD has largely made planning redundant and views can be scaled and moved to fit onto the sheet, or the size of the sheet can be increased or reduced as required; however, views are still scaled using standard scales.

## 7.2 Engineering Scales:
A map cannot be of the same size as the area it represents. So, the measurements are scaled down to make the map of a size that can be conveniently used by users such as motorists, cyclists and bushwalkers. A scale drawing of a building (or bridge) has the same shape as the real building (or bridge) that it represents but a different size. Builders use scaled drawings to make buildings and bridges.

In engineering drawing, objects that are drawn to "Full Size" are drawn at a scale of 1:1; i.e. 25 mm on the drawing equals 25 mm on the rule.

Drawing that are reduced are drawn at scales 1:2, 1:5, 1:10, 1:20, 1:50 or 1:100. If a line was measured using a normal 300 mm rule and read 10 mm, then at a scale of 1:2 the dimension would read 20 mm; at scale 1:5 the dimension would read 100 mm; at scale 1:10 the dimension would read 200 mm; at scale 1:20 the dimension would read 400 mm; at scale 1:50 the dimension would read 1000 mm while at scale 1:100 the dimension would read 2000 mm.

Drawing that are enlarged are drawn at scales 2:1, 5: and 10:1. If a line was measured using a normal 300 mm rule and read 200 mm, then at a scale of 2:1 the dimension would read 100 mm; at scale 5:1 the dimension would read 40 mm while at scale 10:1 the dimension would read 20 mm.

# MEM09002B – Interpret technical drawing
## Topic 7 - Scales

Reading the standard mechanical engineer's scale rules are rather simple because the scale is plainly marked with minor graduations representing 1 mm, 2 mm 5, mm 10, mm, 20 mm, 25 mm, 50 mm and 100 mm. Major graduations are marked in rounded off numbers according to the indicated scale.

### 7.2.1 Imperial Scales:
The imperial scale system is based in the English (Foot-Inch) system of measurement. The imperial system of measurement was reintroduced to England in 1066 and is based on the old Roman measures. The units were standardized in 1215 and updated in 1496, 1588, 1758 and 1824. The primary units of measurement used in the imperial system are inches (in or ") and feet (ft or '); there are 12-inches to the foot and 3-feet to the yard. The inches are divided into fractions; halves, quarters, eights, sixteenths, thirty-seconds and sixty-fourths.

The preferred imperial scales are 1"=1', ¾"=1', ½"=1', ⅜"=1', ¼"=1' and ⅛"=1'.

### 7.2.2 Metric Scales:
The metric system uses the "meter" as the standard for linear measurement. The meter was established by the French in 1791 and its length represents one ten-millionth of the distance from the earth's equator to the pole. The primary unit of measurement used in engineering, civil, architectural, and most other drafting disciplines is the millimeter (mm); secondary units of measurement are the meter (m) and kilometer (km). In Australia, the metric system for all measurements on engineering drawings was commenced in 1975.

The preferred metric scales used are: 1:1, 1:2, 1:5, 1:10, 1:20, 1:50, 1:100, 1:200, and 1:500; other larger scales are obtained by multiplying by a factor of 10. Alternate scales of 1:2.5, 1:25, and 1:250 can also be used if absolutely required.

The metric system has been developed based on scientific principles to meet the requirements of science and trade where the imperial system evolved without any constraints apart from a person's body; the metric system offers a number of substantial advantages:
- Simplicity – The metric system has 7 basic measures while the imperial system has over 300.
- Ease of Calculation – All the units in the metric system are multiplied or divided by 10 to make larger or smaller units.
- International Standard – All major countries except the U.S.A. have converted to the metric system.

## 7.3 Reading Scale Rules:
All scale rules are divided into equal divisions, the distance varying depending on the scale being used. The smaller marks on 1:1 scales represent 1mm while the same division represents 100mm on a 1:100 scale; the medium divisions read every 5mm, 50mm or 500mm depending on the scale (1:, 1:10 or 1:100). Sometimes it may be necessary to determine a value between the 2 graduations as can be seen in mark "K"; most of the time the value will be half the distance but with practice, an experienced draftsperson can make an accurate guess to 1/10th of the division.

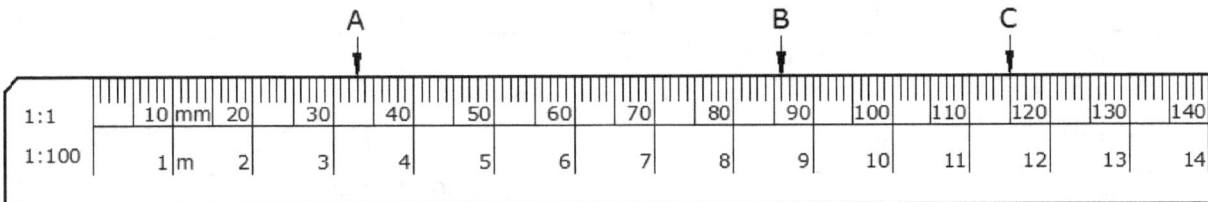

Figure 7.1

To read the 1:10 scale, multiply each graduation by 10 (or place an additional "0" on the top row of values).

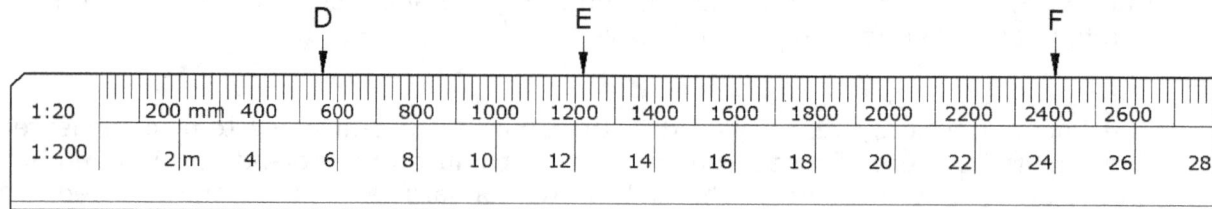

Figure 7.2

Some scale rules may not include the 1:2 scale but instead have 1:20 & 1:200. To read the 1:2 scale, multiply each graduation by 2 (or subtract an "0" on the top row of values).

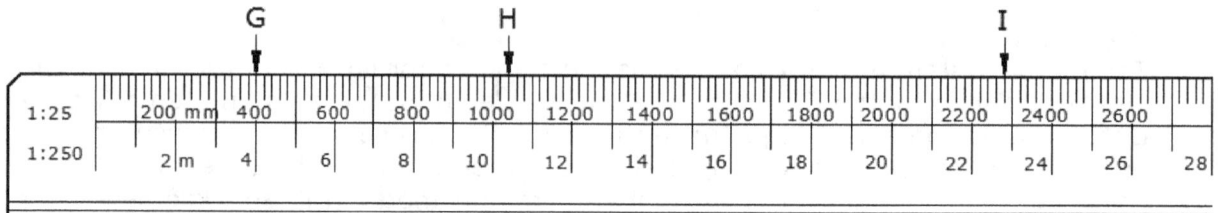

Figure 7.3

To read the 1:2.5 scale, multiply each graduation by 2 (or subtract an "0" on the top row of values).

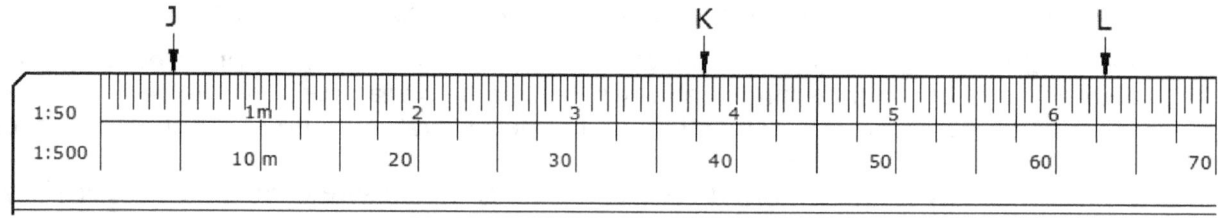

Figure 7.4

To read the 1:5 scale, multiply each graduation by 5 (or imagine the "m" was removed and place two "0"s on the top row of values e.g. 2m reads 200mm).

The following table displays the lengths "as read" in **Error! Reference source not found.**, Figure 7.2, Figure 7.3 & Figure 7.4. All dimensions are in millimetres unless otherwise shown.

| Scale | A | B | C | D | E | F | G | H | I | J | K | L |
|---|---|---|---|---|---|---|---|---|---|---|---|---|
| 1:1 | 33 | 86 | 115 | | | | | | | | | |
| 1:10 | 330 | 860 | 1.15m | | | | | | | | | |
| 1:100 | 3.3m | 8.6m | 11.5m | | | | | | | | | |
| 1:2 | | | | 56 | 122 | 240 | | | | | | |
| 1:20 | | | | 560 | 1220 | 2400 | | | | | | |
| 1:200 | | | | 5.6m | 12.2m | 2.4m | | | | | | |
| 1:2.5 | | | | | | | 40 | 104 | 228 | | | |
| 1:25 | | | | | | | 400 | 1040 | 2280 | | | |
| 1:250 | | | | | | | 4.0m | 10.4m | 22.8m | | | |
| 1:5 | | | | | | | | | | 45 | 380 | 630 |

| 1:50  |  |  |  |  |  |  |  |  |  | 450 | 3800 | 6300 |
|-------|--|--|--|--|--|--|--|--|--|-----|------|------|
| 1:500 |  |  |  |  |  |  |  |  |  | 4.5m | 38.0m | 63.0m |

## 7.4 Recommended Scales:

| Enlargement | Full Size | Reduction | |
|---|---|---|---|
| 10:1 | 1:1 | 1:2 | 1:20 |
| 5:1 |  | 1:2.5 | 1:25 |
| 2:1 |  | 1:5 | 1:50 |
|  |  | 1:10 | 1:100 |

From the various scaled views in Figure 7.5 it can be seen that the larger the scale the more detail is shown however some views may become too large for the drawing and the part may need to be broken into a part section or detail view as shown in the Undercut Detail at Scale 5:1. Conversely, if the view is reduced in scale the fine detail cannot be clearly seen and the dimensions are larger than the view as shown in the Shoulder Pin at Scale 1:5.

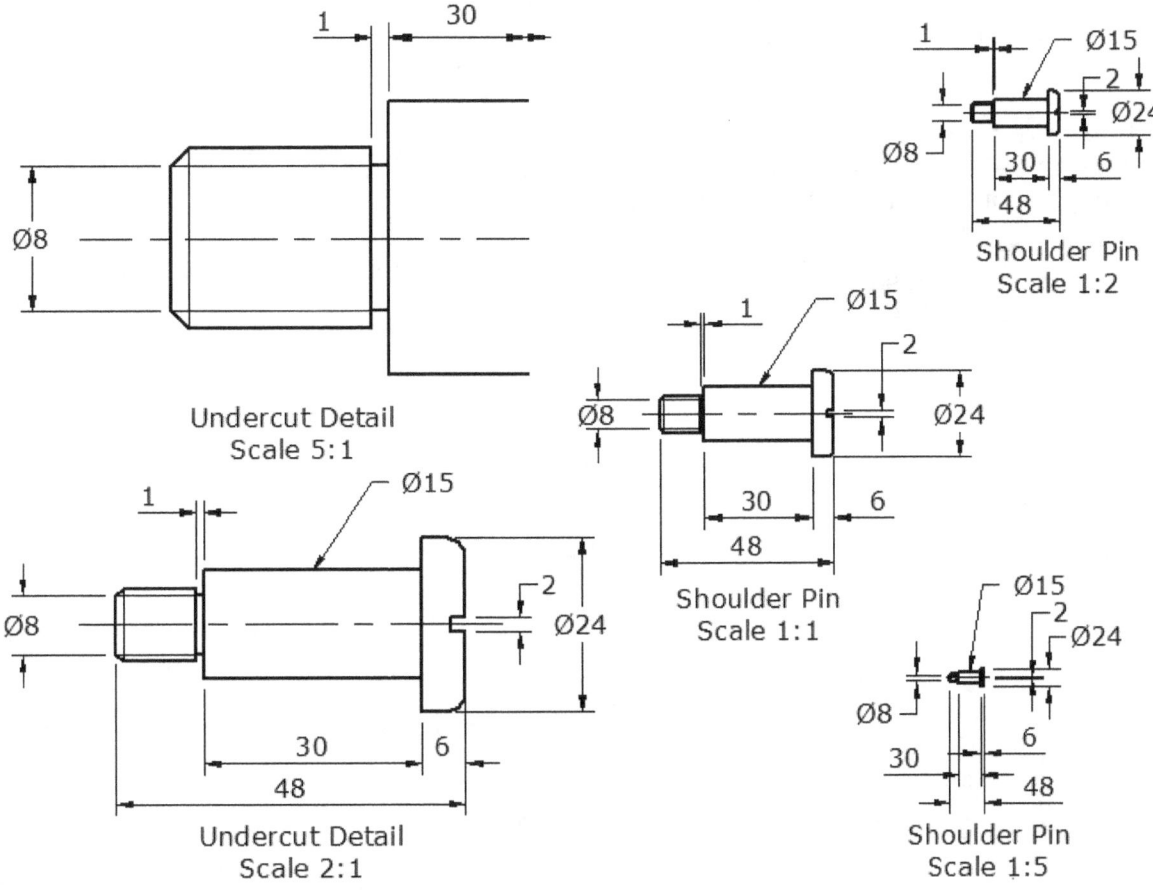

Figure 7.5

## 7.5 Converting Scaled Dimensions to Full Size:

Before expanding on the section a word of warning: The distances on drawing **MUST NEVER** be measured directly from a drawing because during the printing/plotting process, the paper may expand or shrink which will result in inaccurate measurements. Although details are drawn to scale on the original drawing, the hard copy drawing may have been produced on a different size sheet for economic or realistic reasons; e.g. A3 drawings are often plotted on A4 paper due to the availability of A4 printers.

If measurements are to be determined directly from the printed/plotted drawing, those calculations to convert the measured dimension to the full or true size measurement is quite simple and consists of either multiplying or dividing the measured length by the scale.

In the case of reduced scales, the measured length is multiplied by the scale and when enlarged scales are used, the measurement is divided by the scale.

### Example 7-1:

The length of a bracket is measured at 128 mm on a view which is produced at a scale of 1:20, calculate the actual length.

***Procedure:***

Multiply measured length by the scale.

= 128 x 20 = 2560 mm.

### Example 7-2:

The gap between two components is measured at 21.5 mm on a view which is produced at a scale of 5:1, calculate the actual gap.

***Procedure:***

Divide measured gap by the scale.
= 21.5 I 5 = 4.3 mm.

# MEM09002B – Interpret technical drawing
## Topic 7 - Scales

### Skill Practice Exercises:

*Skill Practice Exercise MEM09002-SP-0701*

Determine the scaled measurements to their respective scales from the images below and place the measurements in the table provided.

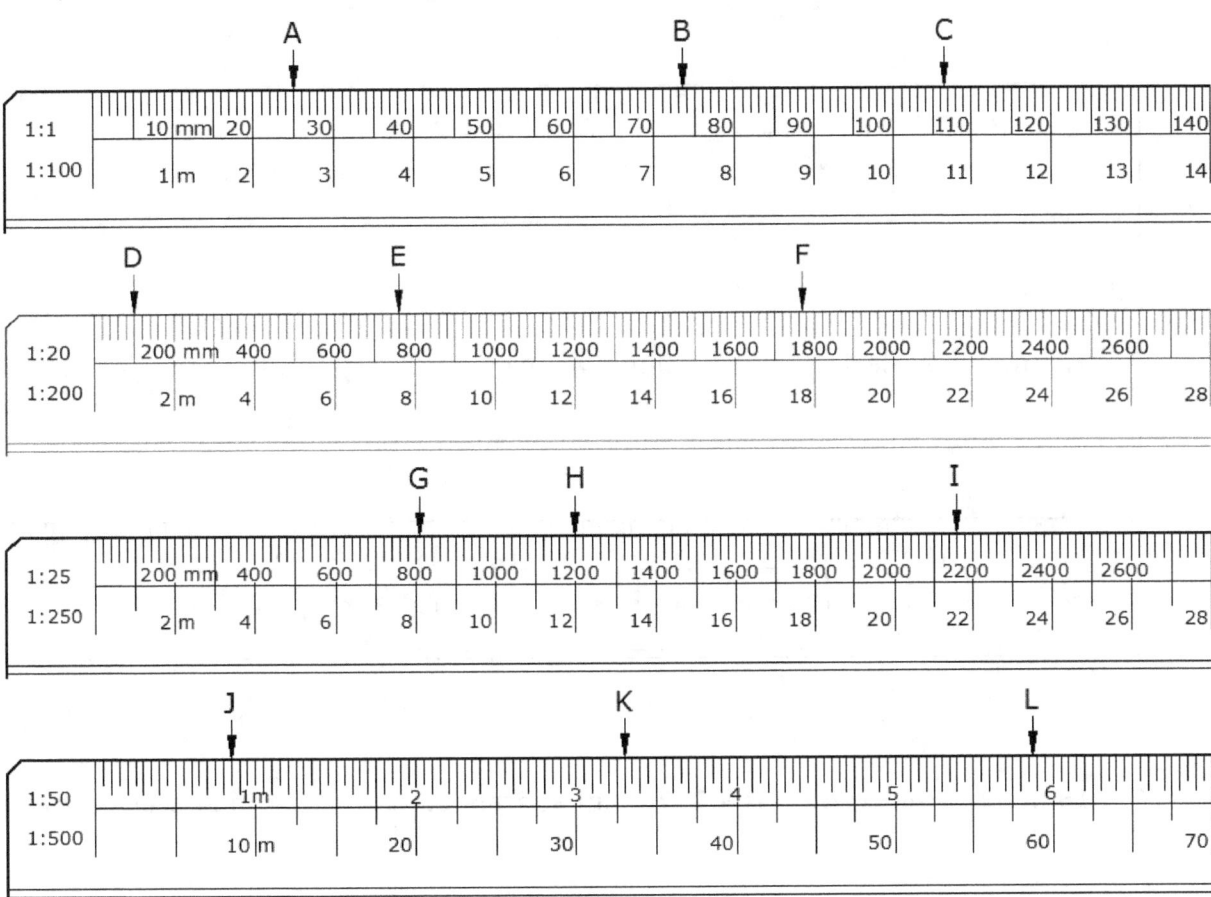

| Scale | A | B | C | D | E | F | G | H | I | J | K | L |
|---|---|---|---|---|---|---|---|---|---|---|---|---|
| 1:1 | | | | | | | | | | | | |
| 1:10 | | | | | | | | | | | | |
| 1:100 | | | | | | | | | | | | |
| 1:2 | | | | | | | | | | | | |
| 1:20 | | | | | | | | | | | | |
| 1:200 | | | | | | | | | | | | |
| 1:2.5 | | | | | | | | | | | | |
| 1:25 | | | | | | | | | | | | |
| 1:250 | | | | | | | | | | | | |
| 1:5 | | | | | | | | | | | | |
| 1:50 | | | | | | | | | | | | |
| 1:500 | | | | | | | | | | | | |

# MEM09002B – Interpret technical drawing
## Topic 7 - Scales

*Skill Practice Exercise MEM09002-SP-0702*
Mark and label the following measurements on the figures of scale rules:

A – 108 mm at scale 1:1    B – 660 mm at scale 1:10    C – 3.55 m at scale 1:100

D – 128 mm at scale 1:2    E – 680 mm at scale 1:20    F – 20.8 m at scale 1:200

G – 18 mm at scale 1:2.5    H – 2070 mm at scale 1:25    I – 13.4 m at scale 1:250

J – 295 mm at scale 1:5    K – 5.7 m at scale 1:50    L – 13.0 m at scale 1:500

**BlackLine Design**
4th October 2015 – Version 3

MEM09002B – Interpret technical drawing

## Topic 7 - Scales

*Skill Practice Exercise MEM09002-SP-0703*
Convert the following scaled measurements to actual sizes:

1. 145.5 @ scale 1:10

2. 75 @ scale 2:1

3. 248 @ scale 1:50

4. 147 @ scale 5:1

5. 298 @ scale 1:1

6. 148.5 @ scale 2:1

7. 635.5 @ scale 1:2

8. 25 @ scale 5:1

9. 67.58 @ scale 1:20

10. 1748.25 @ scale 1:10

# Topic 8 – Abbreviations, Symbols & Notes:

## Required Skills:
- Reading, interpreting information on the drawing.
- Interpret standard abbreviations used in engineering drawings.
- Decipher standard symbols used in engineering drawings.
- Explain standard notes used in engineering drawings.

## Required Knowledge:
- Application of AS1100.
- Understanding of the instructions contained in the drawing.
- Symbols used in the drawing.

## 8.1 Abbreviations & Acronyms:

An abbreviation is any shortened form of a word or phrase and an acronym is a form of an abbreviation; in fact, there are three forms of abbreviation, acronyms, initialism, and truncations.

Abbreviations are used on drawings to save time and space. The common abbreviations such as PCD, THD, MAX, MIN, ID & OD are universally understood. Uncommon abbreviations such as PH BRZ (Phosphor Bronze) should be used cautiously because of the possibility of misinterpretation. Abbreviations should conform to the Australian Standards where possible; non-standard abbreviations should be listed on the drawing. List of abbreviations is given in Appendix 1.

Abbreviations should only be used when their meanings are unquestionably clear to the intended reader; if there is any doubt of confusion, the entire word should be spelt out in full. The same abbreviation should be used for all tenses, the possessive case, participle endings, the singular or plural, and noun and modifying forms.

### 8.1.1 Acronym:
An acronym is a word formed from the initial parts of a name and can consist of letters or syllables. For example, "Australian and New Zealand Army Corps" is commonly known as ANZAC. We are more familiar with sonar than we are with sound navigation and ranging. SONAR is the more recognizable name used in lieu of "Sound Navigation and Ranging".

### 8.1.2 Initialism:
Initialism, or initials, is formed by combining the first letters in a name or expression and each letter is pronounced separately. For example, the "Australian Broadcasting Corporation" is known as the ABC. RFS is the initialism for the Rural Fire Service.

### 8.1.3 Truncation:
In this form of abbreviation, a word is shortened to its first syllable or few letters, for example TUES. is Tuesday and INFO is information.

## 8.2 Jargon:

Jargon is a technical or occupational term developed to help specialists in a specific industry or business communicate quickly and simply with one another. Unfortunately, jargon tends to escape the confines of the narrow fields to which they apply and into our everyday language.

Some examples used in various industries are:

| | | |
|---|---|---|
| Boatbuilding | gravo | A piece of timber or ply used to thicken a weak area on the hull or cabin. |
| | thwart | A seat spanning across a boat, not fore & aft. |
| Construction | foundation | The earth directly beneath the footing and distributes loads to the strata. |
| | stud | A vertical wall member used to attach other structures, such as walls. |
| Fitting & Machining | apron | The portion of a lathe carriage that contains the clutches, gears, and levers for moving the carriage and protects the mechanism. |
| | Bastard File | A medium grade or pitch of file for general purposes, especially suitable for mild steel. |
| Mechanical Engineering | enthalpy | A measure of the total energy of a thermodynamic system. |
| | moment | The tendency of a force to cause a rotation about a point or axis which in turn produces bending stresses. |
| Welding | Back Fire | The momentary burning back of a flame into the tip, followed by a snap or pop, then immediate reappearance or burning out of the flame. |
| | coalescence | The uniting or fusing of metals upon heating. |

## 8.3 Symbols:

Symbols provide a "common language" for drafters all over the world and are intended to communicate design intentions in a clear manner; however, symbols are meaningful only if they are drawn according to relevant standards or conventions. Industry standards have been developed to provide the graphical symbols for the various disciplines. This section describes and illustrates common mechanical, architectural, piping, and electrical symbols.

Unless specified otherwise, the size of dimensioning symbols is consistent with text height. In the majority of cases, all symbols are proportional to the text height.

For a list of symbols used in common drafting disciplines, refer to Appendices 2 to 8.

# MEM09002B – Interpret technical drawing
## Topic 8 - Symbols, Notes & Abbreviations

### Skill Practice Exercises:

*Skill Practice Exercise MEM09002-SP-0801*

Complete the following by adding the abbreviation or the expanded form of the abbreviation.

|  |  |  |  |
|---|---|---|---|
|  | Modulus of Elasticity | FILL HD |  |
|  | Reference | SQ |  |
| CRS |  |  | Specification |
| FREQ |  |  | Rolled Steel Angle |
|  | High Tensile Steel | CHS |  |
| PCD |  |  | Socket |
|  | Galvanised Mild Steel | HYD |  |
| BM |  |  | Hexagonal Head |
|  | Bronze | SPR STL |  |
| UCUT |  |  | Required |
| DIA |  |  | Equivalent |
|  | Miscellaneous | CBORE |  |
|  | Addendum | AL AL |  |
| AP |  |  | Modulus of Inertia |
| CL |  |  | Across Flats |

Name: _____

# MEM09002B – Interpret technical drawing
## Topic 8 - Symbols, Notes & Abbreviations

*Skill Practice Exercise MEM09002-SP-0802*
Complete the following by adding the name or the symbol of the feature.

| Symbol | Name | Symbol | Name |
|---|---|---|---|
| Ø | | (wavy symbol) | Surface must be obtained without machining |
| (bowtie symbol) | | (earth symbol) | |
| | Hanger | (diode symbol) | |
| | Feature Identification | (lightning symbol) | |
| (roof symbol) | | | Loudspeaker |
| | Main Switchboard | (switch symbol) | |
| | All Round Weld | | Circuit Breaker |
| (cross symbol) | | (envelope symbol) | |
| (surface symbol) | | | Slope |
| □ □ | Datum | (antenna symbol) | |
| ⌀ 5 x 40 W | | | Spot Weld |
| | Push Button | | Counterbore |
| ├──┤ | | (target & cone symbols) | |
| ▷ | | (hook symbol) | |
| | Spotlight | | Fillet Weld |
| (V) | | (flag symbol) | |
| | Tunnel Diode | | First Angle Projection |
| | Permanent Magnet | □ | |

Name: _____

# Topic 9 – Reading Drawings:

## Required Skills:
- Checking the drawing against job requirements/related equipment in accordance with standard operating procedures.
- Confirming the drawing version as being current in accordance with standard operating procedures.
- Reading, interpreting information on the drawing.
- Checking and clarifying task related information.

## Required Knowledge:
- Application of AS1100 in accordance with standard operating procedures.
- Relationship between the views contained in the drawing.
- Understanding of the instructions contained in the drawing.
- The materials from which the object(s) are made.

## 9.1 Reading Engineering Drawings:

Engineering drawings are typically used as visual tools in the creation of bridges, towers, airplanes, ships and boats, motor vehicles, gearboxes, braking systems, conveyors, residential industrial and commercial buildings etc.; the range of applications is nearly endless. Anything that is to be fabricated, manufactured or constructed MUST have drawings prepared for planning, estimating, costing, material ordering before any actual the workshop is provided with the drawings for work to commence.

While these drawings can be quite straightforward to individuals who are skilled in the field of engineering or architecture, they can be quite difficult to interpret for laypeople. Knowing how to read engineering drawings will help to provide a better idea of the drawings or plans. The key to interpreting engineering drawings is to understand the purpose of a specific drawing and the relationship of that drawing to the overall set of engineering drawings and specifications prepared for a project.

Reading the orthographic language is a mental process as the drawing is not read aloud unless discussing the drawing with other workers. To describe even a simple object with words is almost impossible.

*"A picture is worth a thousand words."*

Reading proficiency develops with experience, as similar conditions and shapes occur so often that a person in the field gradually acquires a background of knowledge that enables them to visualize readily the shapes shown. Experienced readers read quickly because they can draw upon their knowledge and recognize familiar shapes and combinations without hesitation. However, reading a drawing should always be done carefully and deliberately, as a whole drawing cannot be read at a glance any more than a whole page of print.

### 9.1.1 Prerequisites and Definitions:
Before attempting to read a drawing, familiarize the reader MUST be familiar with the principles of orthographic projection. Keep constantly in mind the arrangement of views and their projection, the space measurements of height, width, and depth, what each line represents, etc.

Visualization is the medium through which the shape information on a drawing is translated to give the reader an understanding of the object represented. The ability to visualize is often thought to be a "gift" that some people possess and others do not; this however, is not true. Any person of reasonable intelligence has a visual memory, as can be seen from their ability to recall and describe scenes at home, actions at sporting events.

The ability to visualize a shape shown on a drawing is almost completely governed by a person's knowledge on the principles of orthographic projection. The common adage that "the best way to learn to read a drawing is to learn how to make one" is quite correct, because in learning to make a drawing one is forced to study and apply the principles of orthographic projection. Reading a drawing can be defined as the process of recognizing and applying the principles of orthographic projection to interpret the shape of an object from the orthographic views.

### 9.1.2 Method of Reading:

A drawing is read by visualizing units or details one at a time from the orthographic projection and mentally orienting and combining these details to interpret the whole object finally. The form taken in this visualization, however, may not be the same for all readers or for all drawings. Reading is primarily a reversal of the process of making drawings; and inasmuch as drawings are usually first made from a picture of the object, the beginner often attempts to carry the reversal too completely back to the pictorial. The result is that the orthographic views of an object like those covered in Topic 5 – Orthographic Projection:

To most, it is a mental impossibility (and surely unnecessary) to translate more than the simplest set of orthographic views into a complete pictorial form that can be pictured in its entirety. Actually, the reader goes through a routine pattern of procedure. Much of this is done subconsciously; for example, consider the object in Figure 9.1. A visible circle is seen in the top view. Memory of previous projection experiences indicates that this must be a hole or the end of a cylinder. The eyes rapidly shift back and forth from the top view to the front view, aligning features of the same size ("in projection"), with the mind assuming the several possibilities and finally accepting the fact that, because of the dashed lines and their extent in the front view, the circle represents a hole that extends through the prism.

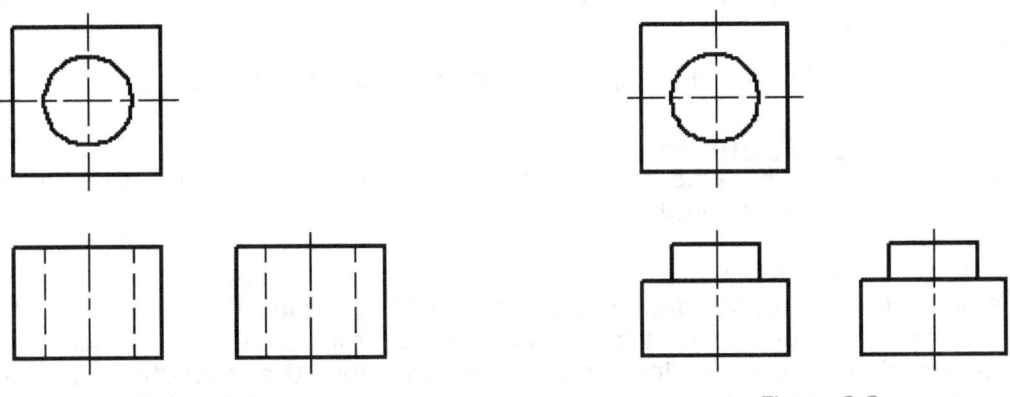

Figure 9.1                                        Figure 9.2

Following a similar pattern of analysis, the reader will find that Figure 9.2 represents a rectangular prism surmounted by a cylinder; this thinking is again done so rapidly, that the reader is scarcely aware of the steps and processes involved.

The following is a suggested technique for reading a drawing; but how does the beginner develop the ability?

*First:*
The reader must have a proficient working knowledge of the principles of orthographic projection.

*Second:*

The reader must acquire a complete understanding of the principles behind the meaning of lines, areas, etc., and the mental process involved in interpreting them, as these principles are applied in reading.

There is very little additional learning required. Careful study of all these items plus practice will develop the ability and confidence needed.

### 9.1.3 Procedure for Reading:

The actual steps in reading are not always identical because of the wide variety of drawings. Nevertheless, the following outline gives the basic procedure and will serve as a guide:

*First:*
The reader should orientate themselves with the views given.

*Second:*
The reader should obtain a general idea of the over-all shape of the object. Think of each view as the object itself, by visualizing being in front, above, and at the side as is done in making the views. Study the dominant features and their relationship to one another.

*Third:*
Start reading the simpler individual features, beginning with the most dominant and progressing to the subordinate. Look for familiar shapes or conditions that your memory retains from previous experience. Read all views of these familiar features to note the extent of holes, thickness of ribs and lugs, etc.

*Fourth:*
Read the unfamiliar or complicated features. Remember that every point, line, surface, and solid appears in every view and that the projection of every detail in the given views must be located to learn the shape.

*Fifth:*
As the reading proceeds, note the relationship between the various portions or elements of the object. Such items as the number and spacing of holes, placement of ribs, tangency of surfaces and the proportions of hubs etc., should be noted and remembered.

*Sixth:*
Reread any detail or relationship that is not clear at the first reading.

## 9.2 Interpreting Drawings:

It is one thing to be able to read a drawing, but that information learned from the drawing must be interpreted.

### 9.2.1 Interpreting a drawing in preparation for manufacture:

It is not usually the prerogative of the designer to decide the details of the machining of a component, although it is often possible to foretell the sequence of some of the manufacturing processes involved; from knowing the manufacturing sequence the designer can identify the manufacturing datum face(s)1, and from this, the required machining dimensions. The datum faces will be those faces used to hold the component during manufacture.

It is common practice in various industries to produce stage drawings. The datum features that are used to produce components stage by stage may not be the same as the finished drawing datum features. For example, a hole may have been produced in the component to allow for the product to line up on a fixture to produce other features, for example, turbine blades slots. Once the slots have then been produced, the hole could be in a feature that is then removed before the component is fully completed. The planning of the manufacturing process based on the stage or final drawings is vital to the success of producing quality component parts. Many areas have to be considered before the manufacturing process begins. Consideration of the following is important:
- What is the required method of manufacture?

- Availability of resources, such as machines, tooling, personnel, equipment.
- How do we hold the component?
- Is fixing required?
- Is in-line measurement used?
- Gauging or measuring instruments?
- What are the best instruments to use?
- Have influences of measurement uncertainty been considered?
- Will training be required?

Once these questions have been answered the production process can begin. It is now important to think about how to monitor the process; our current capability is known, but can the processes consistently be controlled? It may be that as part of the manufacturing process statistical techniques may be used to assist in the interpreting.

**9.2.2 Interpreting a drawing in preparation for measurement:**

The importance of interpreting the design requirements cannot be stressed too highly during preparation for measurement. Identifying the geometric characteristics of the component and the datum features that make up the co-ordinate system is critical to a successful measurement strategy. When making measurements, make use of datum features identified in drawings, technical documents or computer aided design (CAD) models that relate directly to the component.

Datum features on a drawing are normally an important characteristic – a locating or positioning feature. A datum feature could be a face (a surface), a centre line (an axis), or a series of characteristics that collectively make up a datum system. The datum system may be easy to set up when using conventional measuring equipment, such as a surface table in conjunction with angle plates, dial gauges, height gauges and gauge blocks.

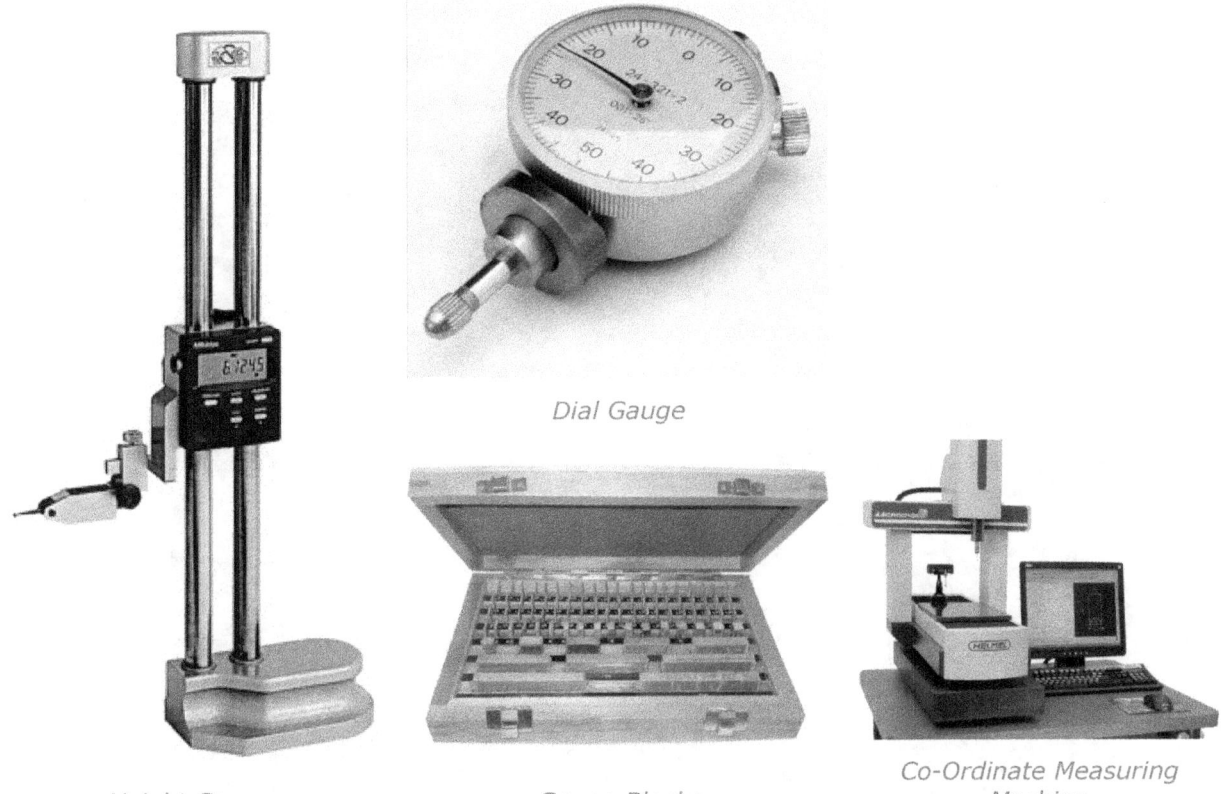

*Dial Gauge*

*Height Gauge*   *Gauge Blocks*   *Co-Ordinate Measuring Machine*

Alternatively, the use of CAD data may be a requirement of the measurement process and, therefore, computerized measuring equipment such as the co-ordinate measuring machine (CMM), may need to be used.

When setting up a datum for measurement it is preferable to choose as datum features the surfaces that were used in the manufacturing process to hold the component; this choice relates the inspection results directly to the manufacturing process.

The features of any component can be defined in two ways, relative to a datum position or positions (absolute), or relative to one another (incremental). The co-ordinate system should be clearly defined whether on a physical drawing or CAD model.

From the drawing the measurement strategy must be determined for the geometric characteristics and the co-ordinate system. Account must be taken of environmental considerations such as temperature effects, equipment required and the associated uncertainties in relation to the stated specifications.

MEM09002B – Interpret technical drawing

Topic 9 - Reading Drawings

## Skill Practice Exercises:

*Skill Practice Exercise MEM09002-RQ-0901*
Refer to drawing STPL-12H-36 and answer the following questions:

1. What grid zone is the Detail of the Oil Grove located?

2. What dimension and grid zone is the dimension drawn "Not to Scale"?

3. What type of section is shown in the right side view?

4. What are the overall dimensions of the Body?

5. How many surfaces are to be machined?

6. Name the method of projection used to produce the drawing.

7. What is the title of the drawing?

8. What size hole must be drilled for the M12x1 tapped hole?

9. What does PCD mean?

10. What is the general size for all unspecified fillets?

11. What is the centre-to-centre distance between the Ø22 holes?

12. What temperature and time must the Body be heated?

13. What is the general tolerance used on the drawing?

14. What is the diameter of the large vertical hole in the Body?

　

15. What is the distance between the toleranced Ø165 and Ø24 holes?

　

16. What change was done to the drawing for Issue C?

　

17. Who checked the drawing?

　

18. What drawing must be referred to obtain details of the Adaptor Plate?

　

19. What scale is used to detail the Oil Groove drawn?

　

20. When are holes to be tapped?

　

21. What is the name of the company producing the drawing?

　

22. What material is used to manufacture the Body?

　

23. Determine the thickness of the base of the Body.

　

24. What radius is used on the Oil Groove?

　

25. What does U.N.O. mean?

　

26. What type of tolerance dimensions are used throughout the drawing?

# MEM09002B – Interpret technical drawing
## Topic 9 - Reading Drawings

*Skill Practice Exercise MEM09002-RQ-0902*

Refer to drawing HC145-58 and answer the following questions:

11. What is the title of the drawing?

___

12. How many pressure gauges are used in the system?

___

13. What is the latest revision number?

___

14. How many Relief Valves are used in the system?

___

15. How many 3 Position 4 Port directional Control Valves are used in the system?

___

16. Who drew the original drawing?

___

17. What is the drawing number?

___

18. Who checked the drawing and date was it checked?

___

19. How many Hydraulic Actuators are used in the system?

___

20. How many Check Valves are used throughout the circuit?

___

21. How many branch points are in the system indicated by dots (•)?

___

22. What change was done to the drawing in Revision C.?

___

23. How many 2 Position 4 Port directional Control Valves are used in the system?

___

24. What does NTS mean?

___

Name: ___

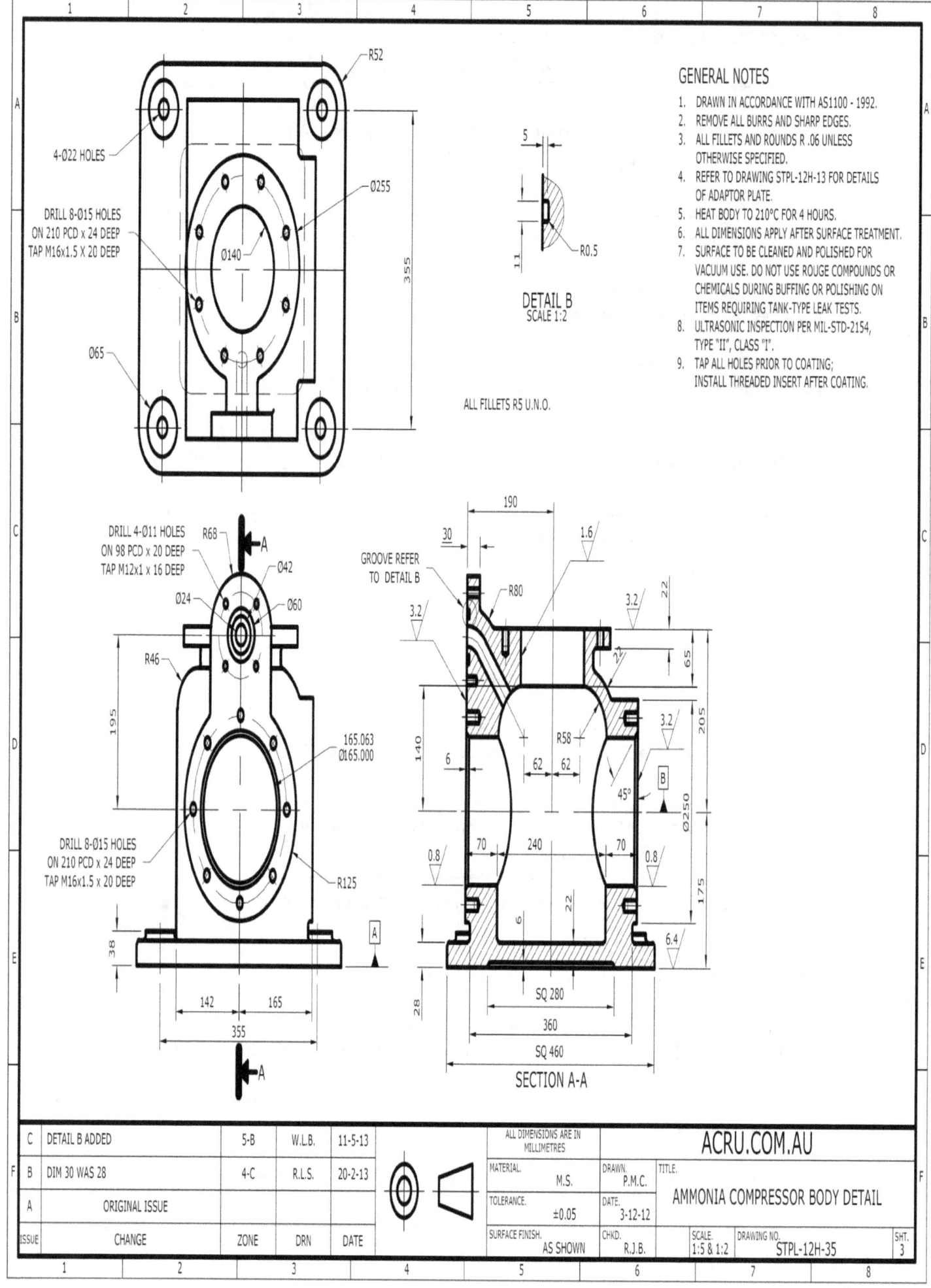

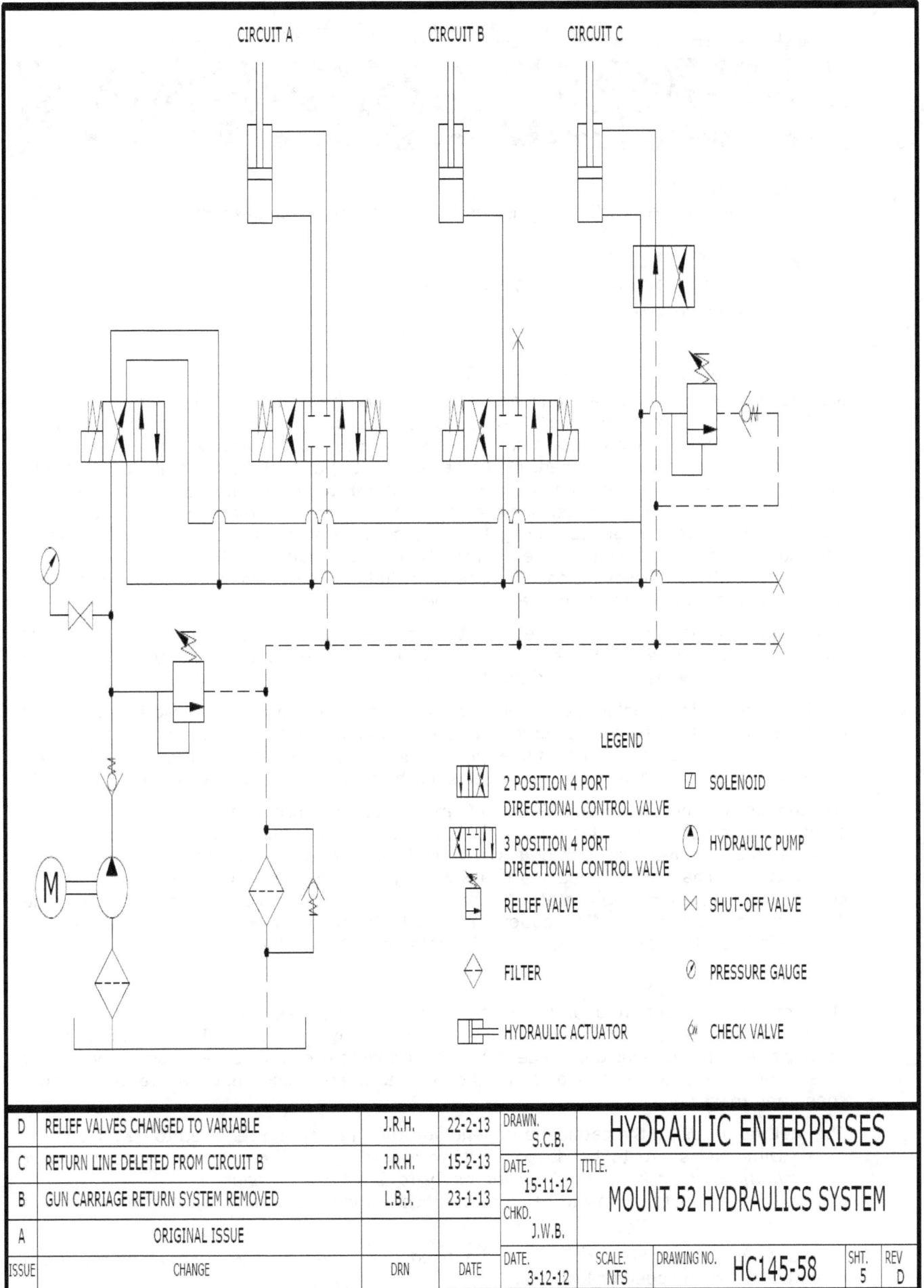

# Topic 10 – Manufacturer's Catalogues:

**Required Skills:**
- Determine specific dimensions from a variety of manufacturer's catalogues.

**Required Knowledge:**
- Reading and Interpreting drawings.
- Measurement.

## 10.1 Introduction:

Manufacturer's catalogues are an important tool used by drafters, technicians, engineers and architects. A manufacturer's catalogue is a document outlining the specifications, performance and other technical characteristics of a product, machine, component (e.g. a pump, electric motor component or structural section) or material, in sufficient detail to be used by a draftsperson or tradesperson to integrate the component into a system. Typically, a datasheet is created by the component/member's manufacturer and begins with an introductory page describing the rest of the document, followed by listings of specific characteristics, with further information on the connectivity of the devices. In cases where there is relevant source code to include, it is usually attached near the end of the document or separated into another file.

Depending on the specific purpose, a data sheet may offer an average value, a typical value, a typical range, engineering tolerances, or a nominal value. The type and source of data are usually stated on the data sheet.

A data sheet is usually used for technical communication to describe technical characteristics of an item or product. It can be published by the manufacturer to help people choose products or to help use the products. By contrast, a technical specification is an explicit set of requirements to be satisfied by a material, product, or service.

Depending on the industry, 99.9% of manufacturers produce catalogues as most components/objects manufactured are different in size and specification from one manufacturer to another; catalogues for windows and doors are used extensively by Architects; Mechanical Engineers are always referring to bearing catalogues while air-conditioning designers need catalogues for A/C units, fans, splitters and ducts etc. The list is nearly unlimited. Catalogues are available normally available in hardcopy or electronic formats with many electronic formats being interactive.

## 10.2 Reading a Catalogue:

Most catalogues contain a draining of the component/assembly with the dimensions comprising letters instead of the numbers. Accompanying the drawing is a table which contains all the dimensions necessary to manufacture the object/assembly; most catalogues provide catalogue and specification numbers, cost, material, ratings, loads, speed and mass etc.

The method for using the catalogue depends on the information required to be asked:
- Dimensions are to be determined from a given part number – *locate the part number and read down the columns for to find the dimension*.
- The part is to fit inside a specific space – *locate the dimension sizes and read across to find the part*.

Variations of the above 2 points can be used depending on the search criteria which could be based on speed or load etc.

# MEM09002B – Interpret technical drawing
## Topic 10 – Manufacturer's Catalogues

*Case Study 1:*

Using the following table for Deep Groove Ball Bearings, determine the designation number and width for a bearing to fit over an Ø35 mm shaft and inside a journal of Ø80 mm.

**Procedure:**

Look down the first column of principle dimensions (d) and locate the 35; it can be seen that the outside diameters in column 2 can be 47,62,72,80 or 100. Reading across the 80 row, the third column indicates the thickness (B) as 21 mm and in the designation column, the bearing number is 6307; the shoulder radius (r) is 1.5 mm.

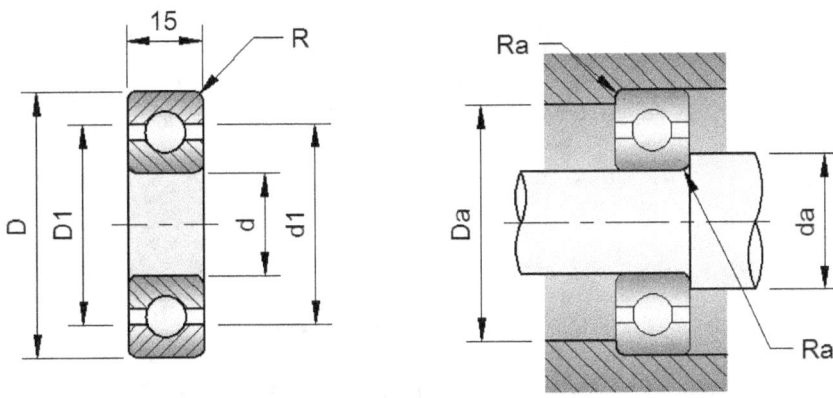

| Principal Dimensions | | | Basic Load Settings | | Limit Speeds | | Mass | Designation | Dimensions | | | | Abutment and Fillet Dimensions | | |
|---|---|---|---|---|---|---|---|---|---|---|---|---|---|---|---|
| | | | | | Lubrication | | | | mm | | | | mm | | |
| d | D | B | C | $C_o$ | Grease | Oil | | | $d_1$ | $D_1$ | $D_2$ | r min | $d_a$ min | $D_a$ max | $r_a$ max |
| mm | | | N | | R/min | | Kg | | mm | | | | mm | | |
| 30 | 52 | 7 | 3120 | 2080 | 15000 | 18000 | 0.026 | 61806 | 33.8 | 38.2 | - | 0.3 | 32 | 40 | 0.3 |
| | 55 | 9 | 11200 | 5850 | 12000 | 15000 | 0.085 | 16006 | 38 | 47.3 | - | 0.3 | 32 | 53 | 0.3 |
| | 55 | 13 | 13300 | 6800 | 12000 | 15000 | 0.12 | 6006 | 38.2 | 47.1 | 49 | 1 | 35 | 50 | 1 |
| | 62 | 16 | 19500 | 10000 | 10000 | 13000 | 0.20 | 6206 | 40.3 | 52.1 | 54.1 | 1 | 35 | 57 | 1 |
| | 72 | 19 | 28100 | 14600 | 9000 | 11000 | 0.35 | 6306 | 44.6 | 59.9 | 61.9 | 1.1 | 36.5 | 65.5 | 1 |
| | 90 | 23 | 43600 | 24000 | 8500 | 10000 | 0.74 | 6406 | 50.3 | 70.7 | - | 1.5 | 38 | 82 | 1.5 |
| 35 | 47 | 7 | 4030 | 3000 | 13000 | 16000 | 0.030 | 61807 | 38.8 | 43.2 | - | 0.3 | 37 | 45 | 0.3 |
| | 62 | 9 | 12400 | 6950 | 10000 | 13000 | 0.11 | 16007 | 44 | 53.3 | - | 0.3 | 37 | 60 | 0.3 |
| | 62 | 14 | 15900 | 8500 | 10000 | 13000 | 0.16 | 6007 | 43.7 | 53.6 | 55.7 | 1 | 40 | 57 | 1 |
| | 72 | 17 | 25500 | 13700 | 9000 | 11000 | 0.29 | 6207 | 46.9 | 60.6 | 62.7 | 1.1 | 41.5 | 65.5 | 1 |
| | 80 | 21 | 33200 | 18000 | 8500 | 10000 | 0.46 | 6307 | 49.5 | 66.1 | 69.2 | 1.5 | 43 | 72 | 1.5 |
| | 100 | 25 | 55300 | 31000 | 7000 | 8500 | 0.95 | 6407 | 57.4 | 80.6 | - | 1.5 | 43 | 92 | 1.5 |
| 40 | 52 | 7 | 4160 | 3350 | 11000 | 14000 | 0.034 | 61808 | 43.8 | 48.2 | - | 0.3 | 42 | 50 | 0.3 |
| | 68 | 9 | 13300 | 7800 | 9500 | 12000 | 0.13 | 16008 | 49.4 | 57 | - | 0.3 | 42 | 66 | 0.3 |
| | 68 | 15 | 16800 | 9300 | 9500 | 12000 | 0.19 | 6008 | 49.2 | 59.1 | 61.1 | 1 | 45 | 63 | 1 |
| | 80 | 18 | 30700 | 16600 | 8500 | 10000 | 0.37 | 6208 | 52.6 | 67.9 | 69.8 | 1.1 | 46.5 | 73.5 | 1 |
| | 90 | 23 | 41000 | 22400 | 7500 | 9000 | 0.63 | 6308 | 56.1 | 74.7 | 77.7 | 1.5 | 48 | 82 | 1.5 |
| | 110 | 27 | 63700 | 36500 | 6700 | 8000 | 1.25 | 6408 | 62.8 | 88 | - | 2 | 49 | 101 | 2 |

# MEM09002B – Interpret technical drawing
## Topic 10 – Manufacturer's Catalogues

*Case Study 2:*
Using the following table for Universal Beams, determine the sizes for a 530UB82.

**Procedure:**
Look down the Designation column and locate 530UB and then the row indicating 82 (or 81.8). Read across the row to determine the sizes, Depth of Section = 528 mm, Flange Width = 209 mm, Flange Thickness = 13.2 mm, Web Thickness = 9.6 mm and Root Radius = 14 mm.

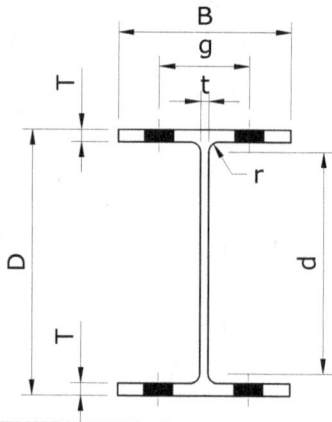

| Designation | Mass Per Metre | Depth of Section | Flange | | Web Thickness | Root Radius | Depth Between Fillets | Gauge Line |
|---|---|---|---|---|---|---|---|---|
| | | | Width | Thickness | | | | |
| | | D | B | T | t | r | d | g |
| | kg | mm | mm | mm | mm | mm | mm | mm |
| 760 UB | 244 | 781 | 272 | 31.3 | 19.3 | 16.5 | 686 | 140 |
| | 220 | 776 | 270 | 28.3 | 17.4 | 16.5 | 686 | |
| | 196 | 770 | 268 | 25.4 | 15.6 | 16.5 | 686 | |
| 690 UB | 140 | 684 | 254 | 19.0 | 12.4 | 15.2 | 615 | 140 |
| | 125 | 678 | 253 | 16.2 | 11.7 | 15.2 | 615 | |
| 610 UB | 125 | 612 | 229 | 19.6 | 11.9 | 14.0 | 572 | 140 |
| | 113 | 607 | 228 | 17.3 | 11.2 | 14.0 | 572 | |
| | 101 | 602 | 228 | 14.8 | 10.6 | 14.0 | 572 | |
| 530 UB | 92.3 | 553 | 209 | 15.6 | 10.2 | 14.0 | 502 | 140 |
| | 81.8 | 528 | 209 | 13.2 | 9.6 | 14.0 | 502 | |
| 460 UB | 81.8 | 460 | 191 | 16.0 | 9.2 | 11.4 | 428 | 90 |
| | 74.4 | 457 | 190 | 14.5 | 9.1 | 11.4 | 428 | |
| | 67.0 | 454 | 190 | 12.7 | 8.5 | 11.4 | 428 | |
| 410 UB | 59.5 | 406 | 178 | 12.8 | 7.8 | 11.4 | 381 | 90 |
| | 53.6 | 403 | 178 | 10.9 | 7.6 | 11.4 | 381 | |
| 360 UB | 56.6 | 359 | 172 | 13.0 | 8.0 | 11.4 | 333 | 90 |
| | 50.6 | 356 | 171 | 11.5 | 7.3 | 11.4 | 333 | |
| | 44.5 | 352 | 171 | 9.7 | 6.9 | 11.4 | 333 | |
| 310 UB | 46.1 | 307 | 166 | 11.8 | 6.7 | 11.4 | 284 | 90 |
| | 40.4 | 304 | 165 | 10.2 | 6.1 | 11.4 | 284 | |
| | 40.2 | 298 | 149 | 8.0 | 5.5 | 13.0 | 282 | |
| 250 UB | 37.2 | 256 | 146 | 10.9 | 6.4 | 8.9 | 234 | 90 |
| | 31.4 | 252 | 146 | 8.6 | 6.1 | | 234 | |
| | 31.3 | 248 | 124 | 8.0 | 5.0 | 12.0 | 232 | |

## 10.3 Data Sheet Index:

Data Sheet 1 – Single Row Deep Groove Ball Bearing:

Data Sheet 2 – Single Row Angular Contact Ball Bearing:

Data Sheet 3 – Tapered Roller Bearings:

Data Sheet 4 – Locking Washers:

Data Sheet 5 – Locking Nuts:

Data Sheet 6 – Felt Sealing Rings:

Data Sheet 7 – Internal Circlips:

Data Sheet 8 – External Circlips:

Data Sheet 9 – Keys and Keyways (Drawing):

Data Sheet 10 – Gibb Head Keys:

Data Sheet 11 – O Rings:

Data Sheet 12 – Parallel Spring Pins:

Data Sheet 13 – Metric Precision Dowel Pins:

Data Sheet 14 – Socket Head Cap Screws:

Data Sheet 15 – Hexagon Socket Set Screws:

Data Sheet 16 – Set Screw Points:

Data Sheet 17 – Structural Steel Sections:

Data Sheet 18 – Universal Beam – Properties & Dimensions:

Data Sheet 19 – Universal Column - Properties & Dimensions:

Data Sheet 20 – Equal Angle - Properties & Dimensions:

Data Sheet 21 – Unequal Angle - Properties & Dimensions:

Data Sheet 22 – Parallel Flange Channel - Properties & Dimensions:

Data Sheet 23 – Electric Motors:

Data Sheet 24 – Rigid Shaft Couplings:

Data Sheet 25 – Speed Reducers:

Data Sheet 26 – Flexible Coupling:

# MEM09002B – Interpret technical drawing
## Topic 10 – Manufacturer's Catalogues

*Data Sheet 1 – Single Row Deep Groove Ball Bearing:*

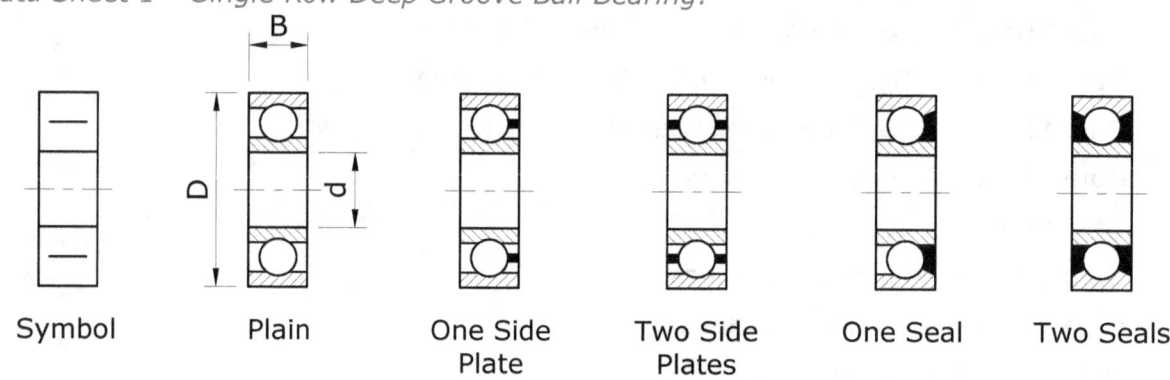

| Symbol | Plain | One Side Plate | Two Side Plates | One Seal | Two Seals |

| Bearing Number | | | | | Dimensions | | | |
|---|---|---|---|---|---|---|---|---|
| Plain | One Side Plate | Two Side Plates | One Seal | Two Seals | d | D | B | r |
| PB-001 | 1SP-001 | 2SP-001 | 1S-001 | 2S-001 | 10 | 26 | 8 | 0.5 |
| PB-002 | 1SP-002 | 2SP-002 | 1S-002 | 2S-002 | 12 | 28 | 8 | 0.5 |
| PB-003 | 1SP-003 | 2SP-003 | 1S-003 | 2S-003 | 15 | 32 | 9 | 0.5 |
| PB-004 | 1SP-004 | 2SP-004 | 1S-004 | 2S-004 | 17 | 35 | 10 | 0.5 |
| PB-005 | 1SP-005 | 2SP-005 | 1S-005 | 2S-005 | 20 | 42 | 12 | 1 |
| PB-006 | 1SP-006 | 2SP-006 | 1S-006 | 2S-006 | 25 | 47 | 12 | 1 |
| PB-007 | 1SP-007 | 2SP-007 | 1S-007 | 2S-007 | 30 | 55 | 13 | 1.5 |
| PB-008 | 1SP-008 | 2SP-008 | 1S-008 | 2S-008 | 35 | 62 | 14 | 1.5 |
| PB-009 | 1SP-009 | 2SP-009 | 1S-009 | 2S-009 | 40 | 68 | 15 | 1.5 |
| PB-010 | 1SP-010 | 2SP-010 | 1S-010 | 2S-010 | 45 | 75 | 16 | 1.5 |
| PB-011 | 1SP-011 | 2SP-011 | 1S-011 | 2S-011 | 50 | 80 | 16 | 1.5 |
| PB-012 | 1SP-012 | 2SP-012 | 1S-012 | 2S-012 | 55 | 90 | 18 | 2 |

**BlackLine Design**
4th October 2015 – Version 3

## Data Sheet 2 – Single Row Angular Contact Ball Bearing:

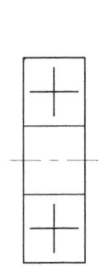
Symbol

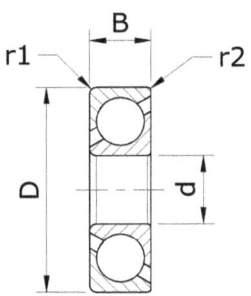

| Bearing No. | Dimensions In millimeters | | | | | Shaft Tols. | Housing Tols. |
|---|---|---|---|---|---|---|---|
| | d | D | B | r1 | r2 | | |
| 72 00 B | 10 | 30 | 9 | 1 | 0.5 | +0.004 / -0.002 | +0.012 / -0.028 |
| 72 01 B | 12 | 32 | 10 | 1 | 0.5 | +0.005 / -0.003 | |
| 72 02 B | 15 | 35 | 11 | 1 | 0.5 | | |
| 72 03 B | 17 | 40 | 12 | 1 | 0.8 | | |
| 72 04 B | 20 | 47 | 14 | 1.5 | 0.8 | +0.005 / -0.004 | +0.014 / -0.033 |
| 72 05 B | 25 | 52 | 15 | 1.5 | 0.8 | | |
| 72 06 B | 30 | 62 | 16 | 1.5 | 0.8 | | |
| 72 07 B | 35 | 72 | 17 | 2 | 1 | +0.006 / -0.005 | |
| 72 08 B | 40 | 80 | 18 | 2 | 1 | | |
| 72 09 B | 45 | 85 | 19 | 2 | 1 | | +0.016 / -0.038 |
| 72 10 B | 50 | 90 | 20 | 2 | 1 | +0.006 / -0.007 | |
| 72 11 B | 55 | 100 | 21 | 2.5 | 1.2 | | |
| 72 12 B | 60 | 110 | 22 | 2.5 | 1.2 | | |
| 72 13 B | 65 | 120 | 23 | 2.5 | 1.2 | | |
| 72 14 B | 70 | 125 | 24 | 2.5 | 1.2 | | |
| 72 15 B | 75 | 130 | 25 | 2.5 | 1.2 | | |

## Data Sheet 3 – Tapered Roller Bearings:

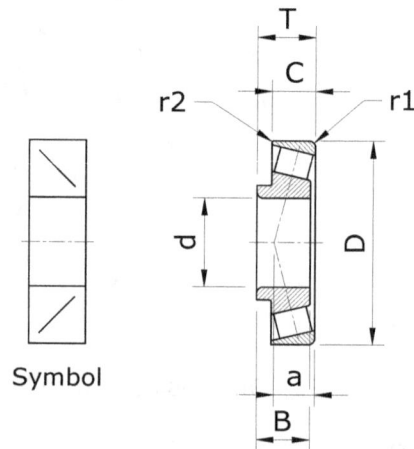

Symbol

| Bearing Number | Dimensions in millimeters | | | | | | | |
|---|---|---|---|---|---|---|---|---|
| | d | D | B | C | T | r | $r_1$ | a |
| TRB - 07 | 35 | 80 | 21 | 18 | 22.75 | 2.5 | 0.8 | 16 |
| TRB -08 | 40 | 90 | 23 | 20 | 25.25 | 2.5 | 0.8 | 19 |
| TRB -09 | 45 | 100 | 25 | 22 | 37.25 | 2.5 | 0.8 | 21 |
| TRB -10 | 50 | 110 | 27 | 23 | 29.25 | 3.0 | 1.0 | 23 |
| TRB -11 | 55 | 120 | 29 | 25 | 31.5 | 3.0 | 1.0 | 24 |
| TRB -12 | 60 | 130 | 31 | 26 | 33.50 | 3.5 | 1.2 | 26 |
| TRB -13 | 65 | 140 | 33 | 28 | 36.00 | 3.5 | 1.2 | 28 |
| TRB -14 | 70 | 150 | 35 | 30 | 38.00 | 3.5 | 1.2 | 29 |
| TRB -15 | 75 | 160 | 37 | 31 | 40.00 | 3.5 | 1.2 | 31 |
| TRB -16 | 80 | 170 | 39 | 33 | 42.50 | 3.5 | 1.2 | 33 |
| TRB -17 | 85 | 180 | 41 | 34 | 44.50 | 4.0 | 1.5 | 35 |
| TRB -18 | 90 | 190 | 43 | 36 | 46.50 | 4.0 | 1.5 | 36 |

*Data Sheet 4 – Locking Washers:*

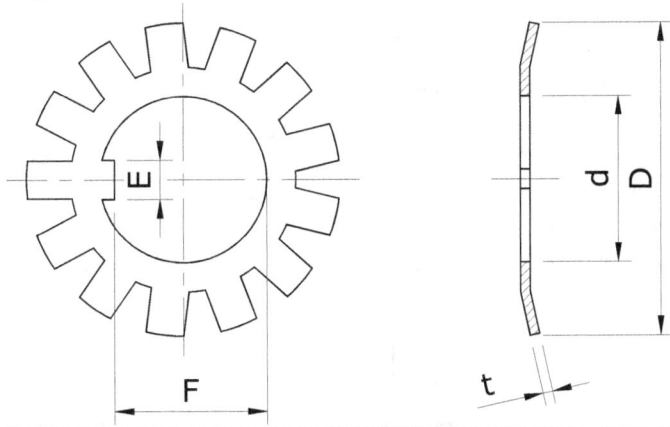

| No. | Dimensions in millimeters | | | | |
|---|---|---|---|---|---|
| | d | D max | t | E | F |
| MB 3 | 17 | 32 | 1 | 4 | 15.5 |
| MB 4 | 20 | 36 | 1 | 4 | 18.5 |
| MB 5 | 25 | 42 | 1.25 | 5 | 23.0 |
| MB 6 | 30 | 49 | 1.25 | 5 | 27.5 |
| MB 7 | 35 | 57 | 1.25 | 6 | 32.5 |
| MB 8 | 40 | 62 | 1.25 | 6 | 37.5 |
| MB 9 | 45 | 69 | 1.25 | 6 | 42.5 |
| MB 10 | 50 | 74 | 1.25 | 6 | 47.5 |
| MB 11 | 55 | 81 | 1.25 | 8 | 52.5 |
| MB 12 | 60 | 86 | 1.5 | 8 | 57.5 |
| MB 13 | 65 | 92 | 1.5 | 8 | 62.5 |
| MB 14 | 70 | 98 | 1.5 | 8 | 66.5 |
| MB 15 | 75 | 104 | 1.5 | 8 | 71.5 |
| MB 16 | 80 | 112 | 1.75 | 10 | 76.5 |
| MB 17 | 85 | 119 | 1.75 | 10 | 81.5 |
| MB 18 | 90 | 126 | 1.75 | 10 | 86.5 |
| MB 19 | 95 | 133 | 1.75 | 10 | 91.5 |
| MB 20 | 100 | 142 | 1.75 | 12 | 96.5 |
| MB 21 | 105 | 145 | 1.75 | 12 | 100.5 |
| MB 22 | 110 | 154 | 1.75 | 12 | 105.5 |

*Data Sheet 5 – Locking Nuts:*

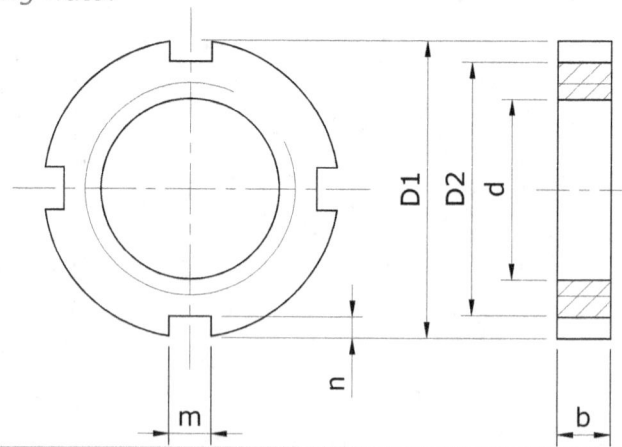

| No. | Dimensions in millimeters | | | | | | Thread | Appertaining Locking Washer |
|---|---|---|---|---|---|---|---|---|
| | d | $D_1$ | $D_2$ | b | m | n | Dia x Pitch | |
| KM 3 | 17 | 28 | 24 | 5 | 4 | 2 | M17 x 1 | MB 3 |
| KM 4 | 20 | 32 | 26 | 6 | 4 | 2 | M20 x 1 | MB 4 |
| KM 5 | 35 | 38 | 32 | 7 | 5 | 2 | M25 x 1.5 | MB 5 |
| KM 6 | 30 | 45 | 38 | 7 | 5 | 2 | M30 x 1.5 | MB 6 |
| KM 7 | 35 | 52 | 44 | 8 | 5 | 2 | M35 x 1.5 | MB 7 |
| KM 8 | 40 | 58 | 50 | 9 | 6 | 2.5 | M40 x 1.5 | MB 8 |
| KM 9 | 45 | 65 | 56 | 10 | 6 | 2.5 | M45 x 1.5 | MB 9 |
| KM 10 | 50 | 70 | 61 | 11 | 6 | 2.5 | M50 x 1.5 | MB 10 |
| KM 11 | 55 | 75 | 67 | 11 | 7 | 3 | M55 x 2 | MB 11 |
| KM 12 | 65 | 80 | 73 | 11 | 7 | 3 | M60 x 2 | MB 12 |
| KM 13 | 65 | 85 | 79 | 12 | 7 | 3 | M65 x 2 | MB 13 |
| KM 14 | 70 | 92 | 85 | 12 | 8 | 3.5 | M70 x 2 | MB 14 |
| KM 15 | 75 | 98 | 90 | 13 | 8 | 3.5 | M75 x 2 | MB 15 |
| KM 16 | 80 | 105 | 95 | 15 | 8 | 3.5 | M80 x 2 | MB 16 |
| KM 17 | 85 | 110 | 102 | 16 | 8 | 3.5 | M85 x 2 | MB 17 |
| KM 18 | 90 | 120 | 108 | 16 | 10 | 4 | M90 x 2 | MB 18 |
| KM 19 | 95 | 125 | 113 | 17 | 10 | 4 | M95 x 2 | MB 19 |
| KM 20 | 100 | 130 | 120 | 18 | 10 | 4 | M100 x 2 | MB 20 |

*Data Sheet 6 – Felt Sealing Rings:*

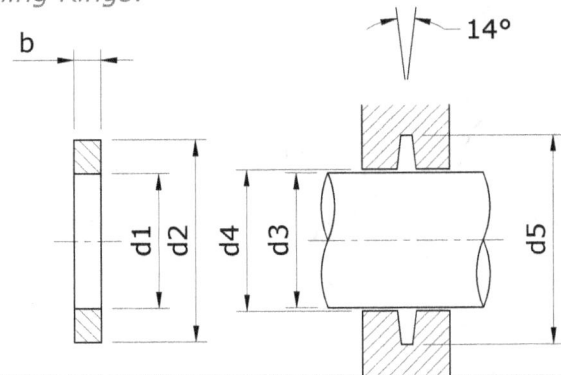

| No. | Dimensions in millimeters | | | | | |
|---|---|---|---|---|---|---|
| | $d_1=d_2$ | b | $d_2$ | $d_4$** | $D_5$ | t |
| Fi 5 | 20 | 4 | 30 | 21 | 31 | 3 |
| Fi 6 | 25 | 5 | 37 | 26 | 38 | 4 |
| Fi 7 | 30 | 5 | 42 | 31 | 43 | 4 |
| Fi 8 | 35 | 5 | 47 | 36 | 48 | 4 |
| Fi 9 | 40 | 5 | 52 | 41 | 53 | 4 |
| Fi 10 | 45 | 5 | 57 | 46 | 58 | 45 |
| Fi 11 | 50 | 6.5 | 66 | 51 | 67 | 5 |
| Fi 12 | 55 | 6.5 | 71 | 56 | 72 | 5 |
| Fi 13 | 60 | 6.5 | 76 | 61.5 | 88 | 5 |
| Fi 15 | 65 | 6.5 | 81 | 66.5 | 82 | 5 |
| Fi 16 | 70 | 7.5 | 88 | 71.5 | 89 | 6 |
| Fi 17 | 75 | 7.5 | 93 | 76.5 | 94 | 6 |
| Fi 18 | 80 | 7.5 | 98 | 81.5 | 99 | 6 |
| Fi 19 | 85 | 7.5 | 103 | 86.5 | 104 | 6 |
| Fi 20 | 90 | 8.5 | 110 | 92 | 111 | 7 |
| Fi 21 | 95 | 8.5 | 115 | 97 | 116 | 7 |

## Data Sheet 7 – Internal Circlips:

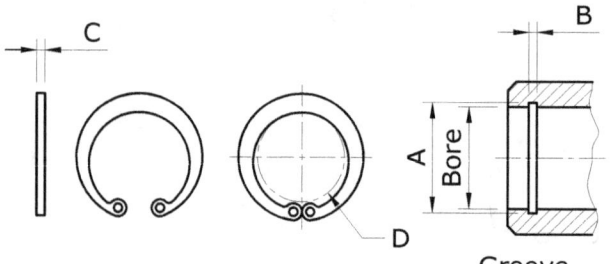
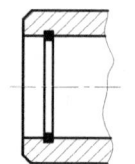

Groove Sizes | Indication on Drawing

Internal Diameter 'D' equals the diameter of the shaft over which the circlip will pass when fitted to bore.

| Bore Dia. | A | B | C | D | Bore Dia. | A | B | C | D |
|---|---|---|---|---|---|---|---|---|---|
| 12 | 12.7 | 1.1 | 1 | 5.5 | 33 | 34.5 | 1.3 | 1.2 | 23 |
| 13 | 13.7 | 1.1 | 1 | 6 | 34 | 35.5 | 1.6 | 1.5 | 24.5 |
| 14 | 14.7 | 1.1 | 1 | 6.75 | 35 | 36.5 | 1.6 | 1.5 | 24.5 |
| 15 | 15.7 | 1.1 | 1 | 8.75 | 36 | 38.5 | 1.6 | 1.5 | 26 |
| 16 | 16.7 | 1.1 | 1 | 8.25 | 38 | 40.2 | 1.6 | 1.5 | 27.5 |
| 17 | 17.9 | 1.1 | 1 | 9 | 40 | 42.5 | 1.85 | 1.75 | 28 |
| 18 | 19 | 1.1 | 1 | 11 | 41 | 43.7 | 1.85 | 1.75 | 30 |
| 19 | 20 | 1.1 | 1 | 10.5 | 42 | 44.7 | 1.85 | 1.75 | 31 |
| 20 | 21.1 | 1.1 | 1 | 11.5 | 44 | 46.7 | 1.85 | 1.75 | 22 |
| 21 | 22.1 | 1.1 | 1 | 11.5 | 45 | 47.7 | 1.85 | 1.75 | 22.5 |
| 22 | 23.1 | 1.1 | 1 | 13 | 46 | 48.8 | 1.85 | 1.75 | 25.5 |
| 23 | 24.1 | 1.1 | 1 | 15 | 47 | 49.8 | 1.85 | 1.75 | 36 |
| 24 | 25.1 | 1.3 | 1.2 | 14 | 48 | 51 | 1.85 | 1.75 | 37 |
| 25 | 26.2 | 1.3 | 1.2 | 16 | 50 | 53 | 2.15 | 2 | 37 |
| 26 | 27.2 | 1.3 | 1.2 | 16 | 51 | 51 | 2.15 | 2 | 37.5 |
| 27 | 28.3 | 1.3 | 1.2 | 17 | 52 | 55 | 2.15 | 2 | 37.5 |
| 28 | 29.3 | 1.3 | 1.2 | 17.5 | 53 | 56.5 | 2.15 | 2 | 40.5 |
| 29 | 30.3 | 1.3 | 1.2 | 18.5 | 54 | 57.5 | 2.15 | 2 | 41.5 |
| 30 | 31.4 | 1.3 | 1.2 | 20 | 55 | 58.5 | 2.15 | 2 | 42 |
| 32 | 33.4 | 1.3 | 1.2 | 22 | 56 | 59.5 | 2.15 | 2 | 43 |

## Data Sheet 8 – External Circlips:

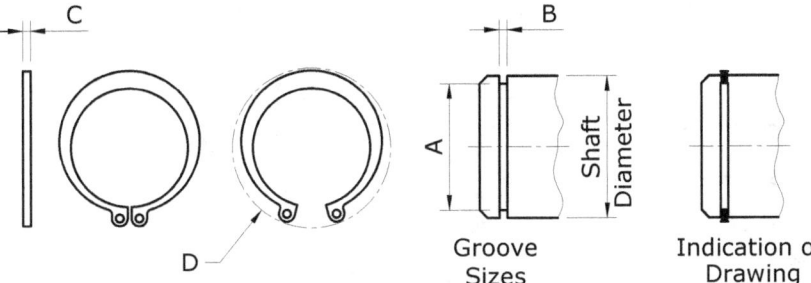

External Diameter 'D' equals the diameter of the shaft over which the circlip will pass when fitted to shaft.

All dimensions are in millimeters.

| Bore Dia. | A | B | C | D | Shaft Dia. | A | B | C | D |
|---|---|---|---|---|---|---|---|---|---|
| 12 | 11.5 | 0.85 | 0.75 | 19.75 | 38 | 36.2 | 1.85 | 1.75 | 51 |
| 13 | 12.5 | 1.1 | 1 | 20.75 | 40 | 38.2 | 1.85 | 1.75 | 51 |
| 14 | 13.4 | 1.1 | 1 | 22.5 | 41 | 39.2 | 1.85 | 1.75 | 53.5 |
| 15 | 14.4 | 1.1 | 1 | 23 | 42 | 40 | 1.85 | 1.75 | 54.5 |
| 16 | 15.3 | 1.1 | 1 | 25.5 | 45 | 43 | 1.85 | 1.75 | 56.5 |
| 17 | 16.3 | 1.1 | 1 | 25 | 47 | 45 | 1.85 | 1.75 | 59 |
| 19 | 18 | 1.3 | 1.2 | 28 | 48 | 46 | 1.85 | 1.75 | 60 |
| 20 | 19 | 1.3 | 1.2 | 28.5 | 50 | 47.8 | 2.15 | 2 | 63 |
| 21 | 20 | 1.3 | 1.2 | 29.5 | 51 | 48.8 | 2.15 | 2 | 63.5 |
| 22 | 21 | 1.3 | 1.2 | 31.5 | 54 | 51.8 | 2.15 | 2 | 66 |
| 24 | 23 | 1.3 | 1.2 | 32 | 55 | 52.8 | 2.15 | 2 | 67 |
| 25 | 24 | 1.3 | 1.2 | 34.5 | 57 | 54.8 | 2.15 | 2 | 72 |
| 26 | 25 | 1.3 | 1.2 | 37 | 60 | 57.5 | 2.15 | 2 | 72 |
| 27 | 26 | 1.3 | 1.2 | 37.5 | 63 | 60.5 | 2.15 | 2 | 75.5 |
| 28 | 27 | 1.6 | 1.5 | 40 | 64 | 61.5 | 2.15 | 2 | 76.5 |
| 29 | 27.5 | 1.6 | 1.5 | 40.5 | 65 | 62.5 | 2.65 | 2.5 | 79 |
| 30 | 28.5 | 1.6 | 1.5 | 40.5 | 70 | 67.5 | 2.65 | 2.5 | 84 |
| 32 | 30.5 | 1.6 | 1.5 | 43.5 | 75 | 72.3 | 2.65 | 2.5 | 91 |
| 34 | 32.5 | 1.6 | 1.5 | 45.5 | 76 | 73.3 | 2.65 | 2.5 | 92 |
| 35 | 33.5 | 1.6 | 1.5 | 47.5 | 80 | 77.3 | 2.65 | 2.5 | 94 |

## Data Sheet 9 – Keys and Keyways (Drawing):

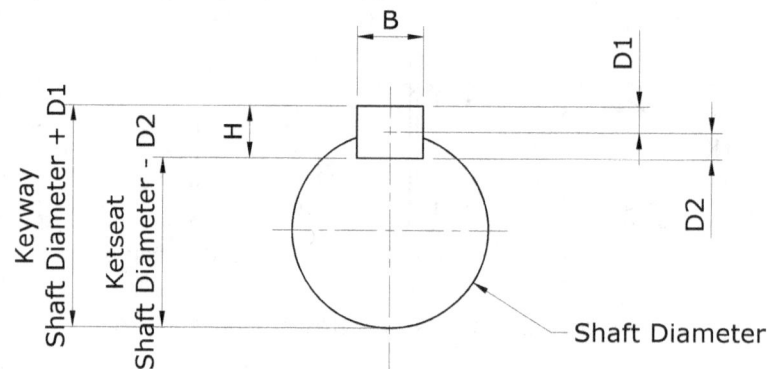

| Shaft Diameter | | Key Section B x H | Width B | | | | | Depth | | | |
|---|---|---|---|---|---|---|---|---|---|---|---|
| Over | Up To & Inc | | Sliding Key | | Normal Key | | Fitted Key | Shaft D₂ | | Hub D₁ | |
| | | | Shaft Tol. | Hub Tol. | Shaft Tol. | Hub Tol. | Shaft & Hub Tol. | Nom. | To. | Nom. | Tol. |
| 6 | 8 | 2x2 | +0.025 0 | +0.060 +0.020 | -0.004 -0.029 | +0.013 -0.013 | -0.006 -0.031 | 1.2 | +0.1 0 | 1 | +0.1 0 |
| 8 | 10 | 3x3 | | | | | | 1.8 | | 1.4 | |
| 10 | 12 | 4x4 | +0.030 0 | +0.078 +0.030 | 0 -0.036 | +0.015 -0.015 | -0.012 -0.042 | 2.5 | | 1.8 | |
| 12 | 17 | 5x5 | | | | | | 3 | | 2.3 | |
| 17 | 22 | 6x6 | | | | | | 3.5 | | 2.8 | |
| 22 | 30 | 8x7 | +0.036 0 | +0.089 +0.040 | 0 -0.036 | +0.018 -0.018 | -0.015 -0.051 | 4 | | 3.3 | |
| 30 | 38 | 10x8 | | | | | | 5 | | 3.3 | |
| 38 | 44 | 12x8 | +0.043 0 | +0.120 +0.060 | 0 -0.043 | +0.022 -0.022 | -0.018 -0.0616 | 5 | +0.2 0 | 3.3 | +0.2 0 |
| 44 | 60 | 14x9 | | | | | | 5.5 | | 3.8 | |
| 50 | 58 | 16x10 | | | | | | 6 | | 4.3 | |
| 58 | 65 | 18 x11 | | | | | | 7 | | 4.4 | |
| 65 | 75 | 20x12 | +0.052 0 | +0.149 +0.065 | 0 -0.052 | +0.026 -0.026 | -0.022 -0.074 | 7.5 | | 4.9 | |
| 75 | 85 | 22x14 | | | | | | 9 | | 5.4 | |
| 85 | 95 | 25x14 | | | | | | 9 | | 5.4 | |
| 95 | 110 | 28x16 | | | | | | 10 | | 6.4 | |
| 110 | 130 | 32x18 | +0.062 0 | +0.180 +0.080 | 0 -0.062 | +0.031 -0.031 | -0.026 -0.088 | 11 | | 7.4 | |
| 130 | 150 | 36x20 | | | | | | 12 | | 8.4 | |
| 150 | 170 | 40x22 | | | | | | 13 | | 9.4 | |
| 170 | 200 | 45x25 | | | | | | 15 | | 10.4 | |
| 200 | 230 | 50x28 | | | | | | 17 | +0.3 0 | 11.4 | +0.3 0 |
| 230 | 260 | 56x32 | +0.074 0 | +0.022 +0.100 | 0 -0.074 | +0.037 -0.037 | -0.032 -0.106 | 20 | | 12.4 | |
| 260 | 290 | 63x32 | | | | | | 20 | | 12.4 | |
| 290 | 330 | 70x36 | | | | | | 22 | | 14.4 | |
| 330 | 380 | 80x40 | | | | | | 25 | | 15.4 | |
| 380 | 440 | 90x45 | +0.087 0 | +0.260 +0.120 | 0 -0.087 | +0.044 -0.044 | -0.037 -0.124 | 28 | | 17.4 | |
| 440 | 500 | 100x50 | | | | | | 31 | | 19.5 | |

## Data Sheet 10 – Gibb Head Keys:

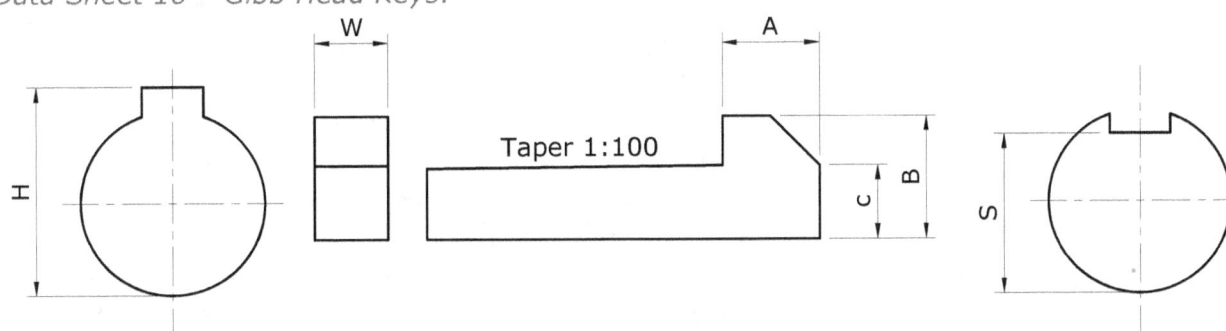

H = Shaft Diameter + X                                    S = Shaft Diameter

| Shaft | | Nom. | Nom | Gibb Head | | | Key Depth | |
|---|---|---|---|---|---|---|---|---|
| Over | To | W | T | A | B | C | X | Y |
| 12 | 20 | 5 | 4 | 6 | 9 | 5 | 2.37 | 1.56 |
| 20 | 25 | 6 | 5 | 8 | 10 | 6 | 2.87 | 1.84 |
| 25 | 32 | 8 | 6 | 10 | 12 | 7 | 3.35 | 2.13 |
| 32 | 38 | 10 | 6 | 11 | 13 | 8 | 3.86 | 2.41 |
| 38 | 45 | 11 | 7 | 12 | 15 | 10 | 4.36 | 2.74 |
| 45 | 50 | 12 | 9 | 14 | 17 | 10 | 5.25 | 3.42 |
| 50 | 56 | 14 | 10 | 15 | 20 | 10 | 5.72 | 3.73 |
| 56 | 64 | 15 | 10 | 17 | 21 | 11 | 6.24 | 4.01 |
| 64 | 70 | 16 | 12 | 20 | 23 | 12 | 7.16 | 4.72 |
| 70 | 76 | 20 | 12 | 21 | 28 | 13 | 7.64 | 4.87 |
| 76 | 89 | 22 | 16 | 24 | 29 | 17 | 9.49 | 6.35 |
| 89 | 100 | 25 | 17 | 27 | 33 | 18 | 10.49 | 6.93 |
| 100 | 115 | 29 | 20 | 30 | 36 | 20 | 11.48 | 7.51 |
| 115 | 127 | 32 | 21 | 33 | 39 | 21 | 12.49 | 8.10 |

## Data Sheet 11 – O Rings:

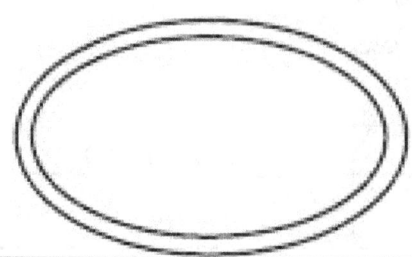

| Code Number | Nominal ID | Section Dia. | Code Number | Nominal ID | Section Dia. | Code Number | Nominal ID | Section Dia. |
|---|---|---|---|---|---|---|---|---|
| 1 | 3 | 1.5 | 23 | 32 | 3.0 | 45 | 92 | 5.0 |
| 2 | 4 | 1.5 | 24 | 33 | 3.0 | 46 | 95 | 5.0 |
| 3 | 5 | 1.5 | 25 | 35 | 3.0 | 47 | 98 | 5.0 |
| 4 | 5.5 | 1.5 | 26 | 37 | 3.0 | 48 | 100 | 5.0 |
| 5 | 6 | 1.5 | 27 | 38 | 3.0 | 49 | 104 | 5.0 |
| 6 | 8 | 1.5 | 28 | 38 | 5.0 | 50 | 108 | 5.0 |
| 7 | 10 | 1.5 | 29 | 41 | 5.0 | 51 | 111 | 5.0 |
| 8 | 10 | 2.0 | 30 | 44 | 5.0 | 52 | 114 | 5.0 |
| 9 | 11 | 2.0 | 31 | 48 | 5.0 | 53 | 117 | 6.0 |
| 10 | 12 | 2.0 | 32 | 50 | 5.0 | 54 | 120 | 6.0 |
| 11 | 14 | 2.0 | 33 | 54 | 5.0 | 55 | 124 | 6.0 |
| 12 | 16 | 2.0 | 34 | 57 | 5.0 | 56 | 127 | 6.0 |
| 13 | 17 | 2.0 | 35 | 60 | 5.0 | 57 | 130 | 6.0 |
| 14 | 19 | 2.0 | 36 | 64 | 5.0 | 58 | 133 | 6.0 |
| 15 | 19 | 3.0 | 37 | 67 | 5.0 | 59 | 136 | 6.0 |
| 16 | 20 | 3.0 | 38 | 70 | 5.0 | 60 | 140 | 6.0 |
| 17 | 22 | 3.0 | 39 | 73 | 5.0 | 61 | 143 | 6.0 |
| 18 | 24 | 3.0 | 40 | 75 | 5.0 | 62 | 146 | 6.0 |
| 19 | 25 | 3.0 | 41 | 79 | 5.0 | 63 | 149 | 6.0 |
| 20 | 27 | 3.0 | 42 | 83 | 5.0 | 64 | 150 | 6.0 |
| 21 | 29 | 3.0 | 43 | 86 | 5.0 | 65 | 159 | 6.0 |
| 22 | 30 | 3.0 | 44 | 89 | 5.0 | 66 | 165 | 6.0 |

## MEM09002B – Interpret technical drawing
### Topic 10 – Manufacturer's Catalogues

*Data Sheet 12 – Parallel Spring Pins:*

| Length | Diameter | | | | | | | | | | | |
|---|---|---|---|---|---|---|---|---|---|---|---|---|
| | 1.5 | 2 | 2.5 | 3 | 4 | 5 | 5.5 | 6 | 8 | 10 | 11 | 12 |
| 5 | 5-1 | 5-2 | | | | | | | | | | |
| 6 | 6-1 | 6-2 | 6-3 | 6-4 | | | | | | | | |
| 8 | 8-1 | 8-2 | 8-3 | | | 8-4 | | | | | | |
| 10 | 10-1 | 10-2 | 10-3 | 10-4 | | | | | | | | |
| 11 | 11-1 | 11-2 | 11-3 | 11-4 | 11-5 | | | | | | | |
| 12 | 12-1 | 12-2 | 12-3 | 12-4 | 12-5 | | 12-6 | | | | | |
| 14 | 14-1 | 14-2 | 14-3 | 14-4 | 14-5 | | 14-6 | | | | | |
| 16 | 16-1 | 16-2 | 16-3 | 16-4 | 16-5 | 16-5 | 16-7 | 16-8 | | | | |
| 18 | 18-1 | 18-2 | 18-3 | 18-4 | 18-5 | 18-6 | 18-7 | 18-8 | | | | |
| 20 | 20-1 | 20-2 | 20-3 | 20-4 | 20-5 | 20-6 | 20-7 | 20-8 | 20-9 | 20-10 | | |
| 21 | | 21-1 | 21-2 | 21-3 | 21-4 | 21-5 | | 21-6 | 21-7 | | | |
| 22 | 22-1 | 22-2 | 22-3 | 22-4 | 22-5 | 22-6 | 22-7 | 22-8 | | | | |
| 24 | | | | 24-1 | 24-2 | 24-3 | 24-4 | 24-5 | 24-6 | 24-7 | 24-8 | |
| 25 | 25-1 | 25-2 | 25-3 | 25-4 | 25-5 | 25-6 | 25-7 | 25-8 | 25-9 | | | |
| 32 | | 32-1 | 32-2 | 32-3 | 32-4 | 32-5 | 32-6 | 32-7 | 32-8 | 32-9 | 32-10 | 32-11 |
| 35 | | | | 35-1 | 35-2 | 35-3 | 35-4 | 35-5 | 35-6 | | | |
| 38 | | 38-1 | 38-2 | 38-3 | 38-4 | 38-5 | 38-6 | 38-7 | 38-8 | 38-9 | 38-10 | 38-12 |
| 41 | | | | 41-1 | 41-2 | 41-3 | 41-4 | 41-5 | 41-6 | 41-7 | 41-8 | 41-9 |
| 48 | | | | 48-1 | 48-2 | 48-3 | 48-4 | 48-5 | 48-6 | | | |
| 50 | | | | 50-1 | 50-2 | 50-3 | 50-4 | 50-5 | 50-6 | 50-7 | 50-8 | 50-9 |
| 58 | | | | | 58-1 | 58-2 | 58-3 | 58-4 | 58-5 | 58-6 | | 58-7 |
| 64 | | | | | 64-1 | 64-2 | 64-3 | 64-4 | 64-5 | 64-6 | 64-7 | 64-8 |
| 70 | | | | | | | | 70-1 | 70-2 | 70-3 | | 70-4 |
| 75 | | | | | | | 75-1 | 75-2 | 75-3 | 75-4 | 75-5 | 75-6 |
| 89 | | | | | | | 89-1 | 89-2 | 89-3 | 89-4 | | 89-5 |
| 100 | | | | | | | | | 100-1 | 100-2 | | 100-3 |

MEM09002B – Interpret technical drawing

## Topic 10 – Manufacturer's Catalogues

*Data Sheet 13 – Metric Precision Dowel Pins:*

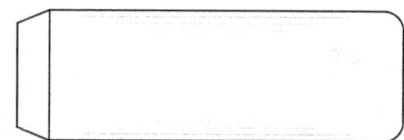

| | Oversize Nominal Size | +0.0010 / +0.0012 |
| --- | --- | --- |
| | Normal Nominal Size | +0.0003 / +0.0001 |

| mm | Nominal Length in millimetres | | | | | | | | | | | | | | | | | | | |
|---|---|---|---|---|---|---|---|---|---|---|---|---|---|---|---|---|---|---|---|---|
| | 4 | 6 | 8 | 10 | 12 | 16 | 20 | 25 | 30 | 35 | 40 | 45 | 50 | 60 | 70 | 80 | 90 | 100 | 110 | 120 |
| 1.5 | ● | ● | ● | ● | ● | | | | | | | | | | | | | | | |
| 2.0 | | ● | ● | ● | ● | ● | ● | | | | | | | | | | | | | |
| 2.5 | | ● | ● | ● | ● | ● | ● | | | | | | | | | | | | | |
| 3 | | | ● | ● | ● | ● | ● | ● | ● | ● | | | | | | | | | | |
| 4 | | | | | ● | ● | ● | ● | ● | ● | ● | | ● | | | | | | | |
| 5 | | | | | | ● | ● | ● | ● | ● | ● | | ● | ● | | | | | | |
| 6 | | | | ● | | ● | ● | ● | ● | ● | ● | ● | ● | | | | | | | |
| 8 | | | | | | ● | ● | ● | ● | ● | ● | ● | ● | ● | | ● | | | | |
| 10 | | | | | | ● | ● | ● | ● | ● | ● | ● | ● | ● | | ● | | ● | | |
| 12 | | | | | | | ● | ● | ● | ● | ● | ● | ● | ● | ● | ● | ● | ● | | |
| 16 | | | | | | | | | | ● | ● | ● | ● | ● | ● | ● | ● | ● | ● | ● |
| 20 | | | | | | | | | | | | ● | ● | ● | ● | ● | ● | ● | ● | ● |
| 25 | | | | | | | | | | | | | | ● | ● | ● | ● | ● | ● | ● |

*Data Sheet 14 – Socket Head Cap Screws:*

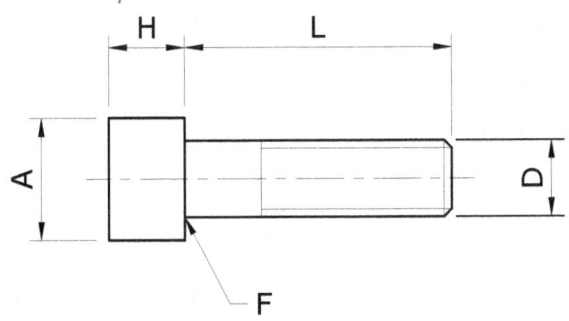

| Nominal Size | Body Diameter D | | Head Diameter A | | Head Depth H | | Hexagon Socket Size | Fillet Radius | |
|---|---|---|---|---|---|---|---|---|---|
| | | | | | | | | F | $d_2$ |
| | Max. | Min. | Max. | Min. | Max. | Min. | Nom. | Max. | Min. |
| M3 | 3.00 | 2.86 | 5.50 | 5.20 | 3.00 | 2.86 | 2.50 | 0.10 | 3.60 |
| M4 | 4.00 | 3.82 | 7.00 | 6.64 | 4.00 | 3.82 | 3.00 | 0.20 | 4.70 |
| M5 | 5.00 | 4.82 | 8.50 | 8.14 | 5.00 | 4.82 | 4.00 | 0.20 | 5.70 |
| M6 | 6.00 | 5.82 | 10.00 | 9.64 | 6.00 | 5.82 | 5.00 | 0.25 | 6.80 |
| M8 | 8.00 | 7.78 | 13.00 | 12.57 | 8.00 | 7.78 | 6.00 | 0.40 | 9.20 |
| M10 | 10.00 | 9.78 | 19.00 | 15.57 | 10.00 | 9.78 | 8.00 | 0.40 | 11.20 |
| M12 | 12.00 | 11.73 | 18.00 | 17.57 | 12.00 | 11.73 | 10.00 | 0.60 | 14.20 |
| M14 | 14.00 | 13.73 | 21.00 | 20.48 | 14.00 | 13.73 | 12.00 | 0.60 | 16.20 |
| M16 | 16.00 | 15.73 | 24.00 | 23.48 | 16.00 | 15.73 | 14.00 | 0.60 | 18.20 |
| M18 | 18.00 | 17.73 | 27.00 | 25.48 | 18.00 | 17.73 | 14.00 | 0.60 | 20.20 |
| M20 | 20.00 | 19.67 | 30.00 | 29.48 | 20.00 | 19.67 | 17.00 | 0.80 | 22.40 |
| M22 | 22.00 | 21.67 | 33.00 | 32.48 | 22.00 | 21.67 | 17.00 | 0.80 | 22.40 |
| M24 | 24.00 | 23.67 | 36.00 | 35.38 | 24.00 | 23.67 | 19.00 | 0.80 | 26.40 |

Data Sheet 15 – Hexagon Socket Set Screws:

 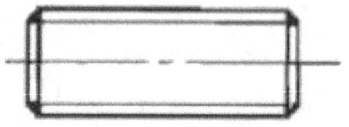

| Length | Diameter in millimetres | | | | | | | | | |
|---|---|---|---|---|---|---|---|---|---|---|
| | M3 | M4 | M5 | M6 | M8 | M10 | M12 | M16 | M20 | M24 |
| 3 | ● | | | | | | | | | |
| 4 | ● | ● | | | | | | | | |
| 5 | ● | ● | ● | | | | | | | |
| 6 | ● | ● | ● | ● | | | | | | |
| 8 | ● | ● | ● | ● | ● | | | | | |
| 10 | ● | ● | ● | ● | ● | ● | | | | |
| 12 | ● | ● | ● | ● | ● | ● | ● | | | |
| 16 | | ● | ● | ● | ● | ● | ● | ● | | |
| 20 | | ● | ● | ● | ● | ● | ● | ● | ● | ● |
| 25 | | | ● | ● | ● | ● | ● | ● | ● | ● |
| 30 | | | ● | ● | ● | ● | ● | ● | ● | ● |
| 35 | | | | ● | ● | ● | ● | | | |
| 40 | | | | ● | ● | ● | ● | ● | ● | ● |
| 50 | | | | | ● | ● | ● | ● | ● | ● |
| 60 | | | | | | | | ● | ● | ● |

# MEM09002B – Interpret technical drawing
## Topic 10 – Manufacturer's Catalogues

*Data Sheet 16 – Set Screw Points:*

| End | Point |
|---|---|
| | Half Dog Point |
| Fluted Socket 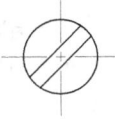 | 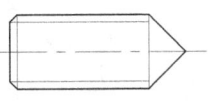 Full Dog Point |
| | Cup Point |
| Slotted | Cone Point |

**BlackLine Design**
4th October 2015 – Version 3

MEM09002B – Interpret technical drawing

## Topic 10 – Manufacturer's Catalogues

*Data Sheet 17 – Structural Steel Sections:*

### Main Structural Steel Sections

| Section Type | Pictorial View | Abbreviation or Symbol | Typical Example of Notation |
|---|---|---|---|
| Universal Beam (depth greater than Flange width) | 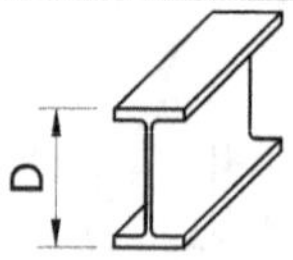 | UB | 200 UB 25<br>Where:<br>200 = D<br>25 = mass, kg/m |
| Universal Column (depth and flange Width approx equal) | 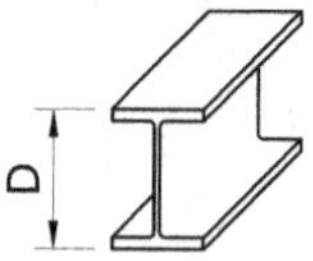 | UC | 200 UC 46<br>Where:<br>200= D<br>46 = mass, kg/m |
| Rolled channel | 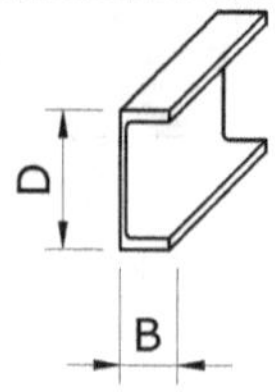 | [ | 76 x 38 x 6.5 [<br>Where:<br>76= D<br>38 = B<br>6.5 = mass, kg/m |
| Rolled Angle (leg width, a, varies may be equal or unequal) | 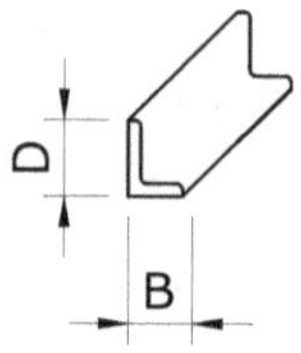 | EA = Equal Angle<br>UA = Unequal Angle | 50 x 50 x 5 UA<br>Where:<br>75= A<br>50 = B<br>5 = T |
| A) Flat Bar or<br>B) Flat Plate<br>(D > 300mm) | 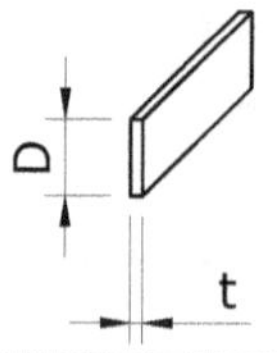 | a) FL<br>b) PL or $\mathcal{P}$ | 200 x 10 PL<br>Where:<br>200= D<br>10 = T |
| Round Bar or Rod | 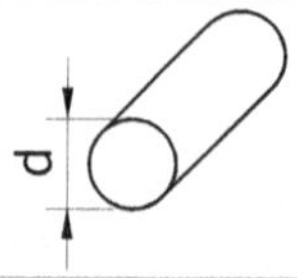 | RD | 16 RD<br>Where:<br>16= D |
| Square Bar | 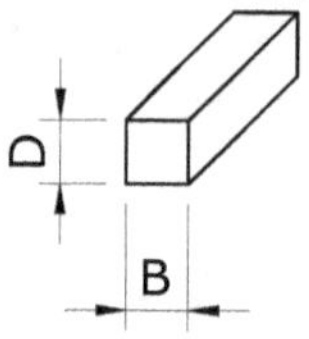 | SQ | 25 SQ<br>Where:<br>25= D or B |
| Circular Hollow Section | 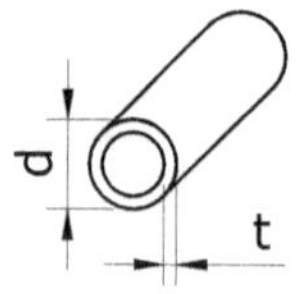 | CHS | 76 OD x 5 CHS<br>Where:<br>76= Outside Diameter<br>5 = t |
| a) Rectangular" Hollow Section<br>b) Square Hollow Section (D = B) | 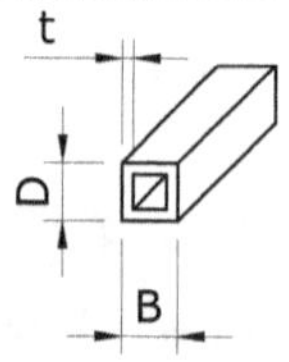 | a) RHS<br>b) SHS | 75 x 50 x 4 RHS<br>Where:<br>75= D<br>50 = B<br>4 = t |

## Detailing Of A Universal Beam

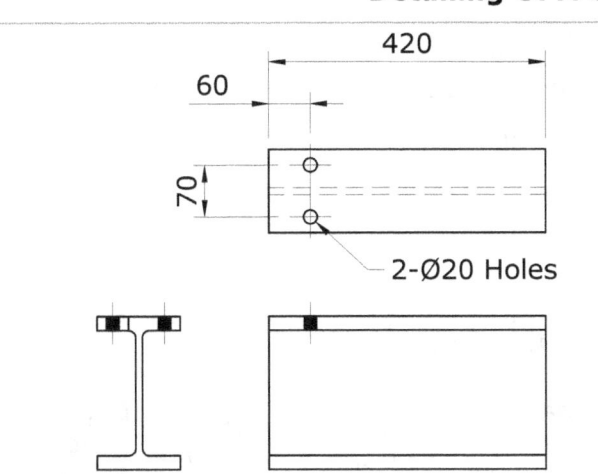

NOTE:

Actual sizes of beam apart from the length are not shown on views

QTY: 3 OFF
MATL : 250 UB x 31.3

## Detailing Of A Channel Section

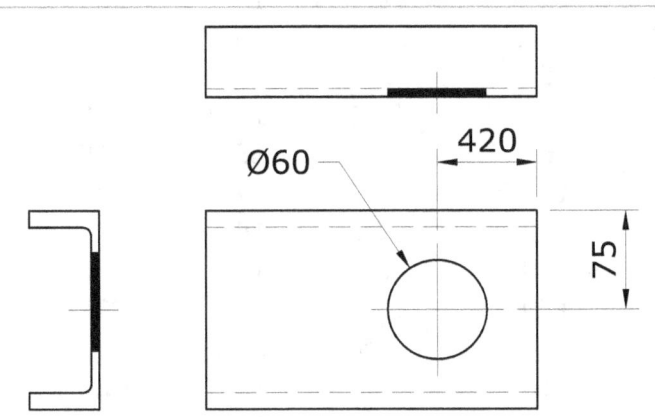

NOTE:
Sizes of channel apart from the length are not shown on views.

Holes through thin sections are shown "filled in".

QTY: 2 OFF
MATL : 127 X 64 X 15 [

## Detailing Of An Angle Section

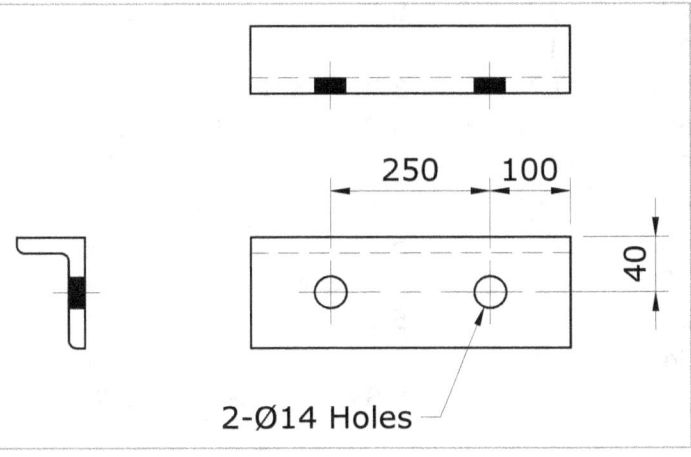

NOTE:

1. Sizes of angles apart from the length are not shown on these views. Dimensions for detailing only are shown

2. Centre lines for holes are marked out from the heel (28) this called the gauge line.

QTY: 40 OFF
MATL : 51 x 51 x 8L

## Data Sheet 18 – Universal Beam – Properties & Dimensions:

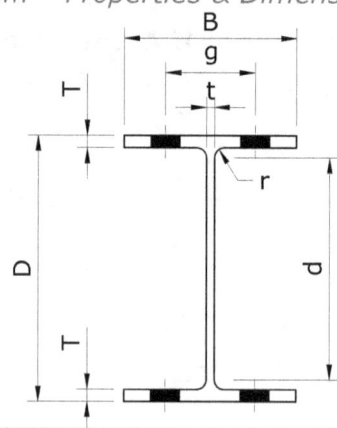

| Designation | Mass Per Metre | Depth of Section | Flange | | Web Thickness | Root Radius | Depth Between Fillets | Gauge Line |
|---|---|---|---|---|---|---|---|---|
| | | | Width | Thickness | | | | |
| | | D | B | T | t | r | d | g |
| | kg | mm | mm | mm | mm | mm | mm | mm |
| 760 UB | 244 | 781 | 272 | 31.3 | 19.3 | 16.5 | 686 | 140 |
| | 220 | 776 | 270 | 28.3 | 17.4 | 16.5 | 686 | |
| | 196 | 770 | 268 | 25.4 | 15.6 | 16.5 | 686 | |
| 690 UB | 140 | 684 | 254 | 19.0 | 12.4 | 15.2 | 615 | 140 |
| | 125 | 678 | 253 | 16.2 | 11.7 | 15.2 | 615 | |
| 610 UB | 125 | 612 | 229 | 19.6 | 11.9 | 14.0 | 572 | 140 |
| | 113 | 607 | 228 | 17.3 | 11.2 | 14.0 | 572 | |
| | 101 | 602 | 228 | 14.8 | 10.6 | 14.0 | 572 | |
| 530 UB | 92.3 | 553 | 209 | 15.6 | 10.2 | 14.0 | 502 | 140 |
| | 81.8 | 528 | 209 | 13.2 | 9.6 | 14.0 | 502 | |
| 460 UB | 81.8 | 460 | 191 | 16.0 | 9.2 | 11.4 | 428 | 90 |
| | 74.4 | 457 | 190 | 14.5 | 9.1 | 11.4 | 428 | |
| | 67.0 | 454 | 190 | 12.7 | 8.5 | 11.4 | 428 | |
| 410 UB | 59.5 | 406 | 178 | 12.8 | 7.8 | 11.4 | 381 | 90 |
| | 53.6 | 403 | 178 | 10.9 | 7.6 | 11.4 | 381 | |
| 360 UB | 56.6 | 359 | 172 | 13.0 | 8.0 | 11.4 | 333 | 90 |
| | 50.6 | 356 | 171 | 11.5 | 7.3 | 11.4 | 333 | |
| | 44.5 | 352 | 171 | 9.7 | 6.9 | 11.4 | 333 | |
| 310 UB | 46.1 | 307 | 166 | 11.8 | 6.7 | 11.4 | 284 | 90 |
| | 40.4 | 304 | 165 | 10.2 | 6.1 | 11.4 | 284 | |
| | 40.2 | 298 | 149 | 8.0 | 5.5 | 13.0 | 282 | |
| 250 UB | 37.2 | 256 | 146 | 10.9 | 6.4 | 8.9 | 234 | 90 |
| | 31.4 | 252 | 146 | 8.6 | 6.1 | | 234 | |
| | 31.3 | 248 | 124 | 8.0 | 5.0 | 12.0 | 232 | |
| 200 UB | 29.8 | 207 | 134 | 9.6 | 6.3 | 8.9 | 188 | 90 |
| | 25.4 | 203 | 133 | 7.8 | 5.8 | 8.9 | 188 | |

## Data Sheet 19 – Universal Column - Properties & Dimensions:

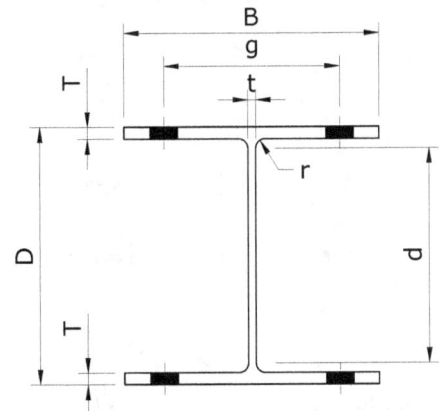

| Designation | Mass Per Metre | Depth of Section | Flange | | Web Thickness | Root Radius | Depth Between Fillets | Gauge Line |
|---|---|---|---|---|---|---|---|---|
| | | | Width | Thickness | | | | |
| | | D | B | T | t | r | d | g |
| | kg | mm | mm | mm | mm | mm | mm | mm |
| 310 UC | 158 | 327 | 311 | 25.0 | 15.7 | 16.5 | 277 | 140 |
| | 137 | 321 | 309 | 21.7 | 13.8 | 16.5 | 277 | |
| | 118 | 315 | 307 | 18.7 | 11.9 | 16.5 | 277 | |
| | 96.8 | 308 | 305 | 15.4 | 9.9 | 16.5 | 277 | |
| 250 UC | 89.5 | 260 | 256 | 17.3 | 10.5 | 14.0 | 225 | 140 |
| | 72.9 | 254 | 254 | 14.2 | 8.6 | 14.0 | 225 | |
| 200 UC | 59.5 | 210 | 205 | 14.2 | 9.3 | 11.4 | 181 | 140 |
| | 52.2 | 206 | 204 | 12.5 | 8.0 | 11.4 | 181 | |
| | 46.2 | 203 | 203 | 11.0 | 7.3 | 11.4 | 181 | |
| 150 UC | 37.2 | 162 | 154 | 11.5 | 8.1 | 8.9 | 139 | 90 |
| | 30.0 | 158 | 153 | 9.4 | 6.6 | 8.9 | 139 | |
| | 23.4 | 152 | 152 | 6.8 | 6.1 | 8.9 | 139 | |
| 100 UC | 14.8 | 97 | 99 | 7.0 | 5.0 | 10.0 | 83 | 70 |

## Data Sheet 20 – Equal Angle - Properties & Dimensions:

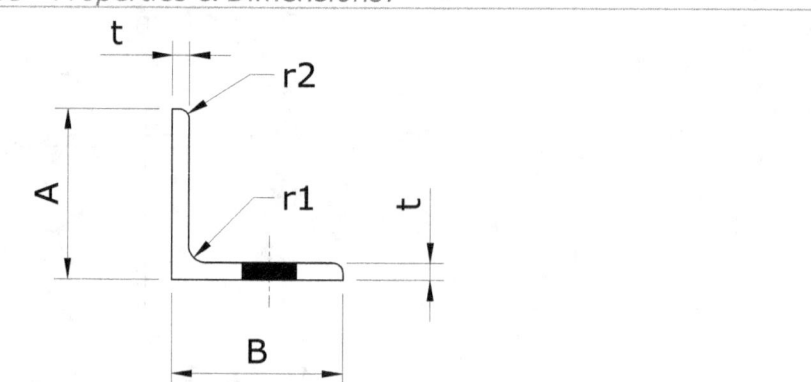

| Designation | Actual Thickness t | Mass | Radii | | Area of Section |
| --- | --- | --- | --- | --- | --- |
| | | | Root r1 | Toe r2 | |
| A x B x t | mm | Kg/m | mm | mm | mm² |
| 75x75x10 EA | 10.5 | 10.5 | 8.00 | 5.00 | 1340 |
| 75x75x8 EA | 8.73 | 8.73 | 8.00 | 5.00 | 1110 |
| 75x75x6 EA | 6.81 | 6.81 | 8.00 | 5.00 | 867 |
| 75x75x5 EA | 5.27 | 5.27 | 8.00 | 5.00 | 672 |
| 65x65x10 EA | 9.02 | 9.02 | 6.00 | 3.00 | 1150 |
| 65x65x8 EA | 7.51 | 7.51 | 6.00 | 3.00 | 957 |
| 65x65x6 EA | 5.87 | 5.87 | 6.00 | 3.00 | 748 |
| 65x65x5 EA | 4.56 | 4.56 | 6.00 | 3.00 | 581 |
| 55x55x6 EA | 6.00 | 4.93 | 6.00 | 3.00 | 628 |
| 55x55x5 EA | 4.60 | 3.84 | 6.00 | 3.00 | 489 |
| 50x50x8 EA | 5.68 | 5.68 | 6.00 | 3.00 | 723 |
| 50x50x6 EA | 4.46 | 4.46 | 6.00 | 3.00 | 568 |
| 50x50x5 EA | 3.48 | 3.48 | 6.00 | 3.00 | 443 |
| 50x50x3 EA | 2.31 | 2.31 | 6.00 | 3.00 | 295 |
| 45x45x6 EA | 3.97 | 3.97 | 5.00 | 3.00 | 506 |
| 45x45x5 EA | 3.10 | 3.10 | 5.00 | 3.00 | 394 |
| 45x45x3 EA | 2.06 | 2.06 | 5.00 | 3.00 | 263 |
| 40x40x6 EA | 3.50 | 3.50 | 5.00 | 3.00 | 446 |
| 40x40x5 EA | 2.73 | 2.73 | 5.00 | 3.00 | 345 |
| 40x40x3 EA | 1.83 | 1.83 | 5.00 | 3.00 | 233 |
| 30x30x6 EA | 2.56 | 2.56 | 5.00 | 3.00 | 326 |
| 30x30x5 EA | 2.01 | 2.01 | 5.00 | 3.00 | 256 |
| 30x30x3 EA | 1.35 | 1.35 | 5.00 | 3.00 | 173 |
| 25x25x6 EA | 2.08 | 2.08 | 5.00 | 3.00 | 266 |
| 25x25x5 EA | 1.65 | 1.65 | 5.00 | 3.00 | 210 |
| 25x25x3 EA | 1.12 | 1.12 | 5.00 | 3.00 | 143 |

*Data Sheet 21 – Unequal Angle - Properties & Dimensions:*

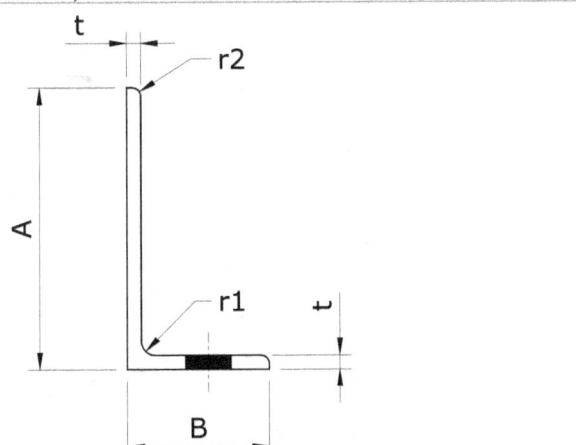

| Designation | Actual Thickness t | Mass | Radii | | Area of Section |
|---|---|---|---|---|---|
| | | | Root | Toe | |
| A x B x t | mm | Kg/m | mm | mm | mm² |
| 100x75x10 UA | 9.5 | 12.4 | 8.0 | 5.0 | 1580 |
| 100x75x8 UA | 7.8 | 10.3 | 8.0 | 5.0 | 1310 |
| 100x75x6 UA | 6.0 | 7.98 | 8.0 | 5.0 | 1020 |
| 75x50x8 UA | 7.8 | 7.23 | 7.0 | 3.0 | 921 |
| 75x50x6 UA | 6.0 | 5.66 | 7.0 | 3.0 | 721 |
| 75x50x5 UA | 4.6 | 4.40 | 7.0 | 3.0 | 560 |
| 65x50x8 UA | 7.8 | 6.59 | 6.0 | 3.0 | 840 |
| 65x50x6 UA | 6.0 | 5.16 | 6.0 | 3.0 | 658 |
| 65x50x5 UA | 4.6 | 4.02 | 6.0 | 3.0 | 512 |

## Data Sheet 22 – Parallel Flange Channel - Properties & Dimensions:

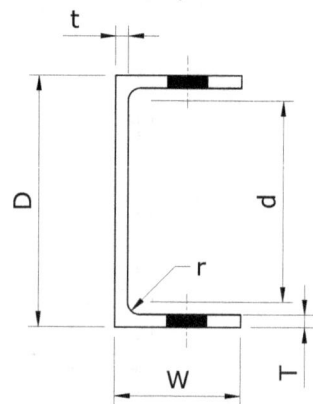

| Desig'n | Mass Per Metre | Depth D | Flange | | Web Thickn's | Root Radius | Depth Between Flanges | Cross Section of Area | Centroid Coord. |
|---|---|---|---|---|---|---|---|---|---|
| | | | Width | Thickn's | | | | | |
| | | D | W | T | t | r | d | | |
| | kg/m | mm | mm | mm | mm | mm | mm | mm² | mm |
| 250 PFC | 35.5 | 250 | 90 | 15.0 | 8.0 | 12.0 | 220 | 4520 | 28.6 |
| 230 PFC | 25.1 | 230 | 75 | 12.0 | 6.5 | 12.0 | 206 | 3200 | 22.6 |
| 200 PFC | 22.9 | 200 | 75 | 12.0 | 6.0 | 12.0 | 176 | 2920 | 24.4 |
| 180 PFC | 20.9 | 180 | 75 | 11.0 | 6.0 | 12.0 | 158 | 2260 | 24.5 |
| 150 PFC | 17.7 | 150 | 75 | 9.5 | 6.0 | 10.0 | 131 | 2250 | 24.9 |
| 125 PFC | 11.9 | 125 | 65 | 7.5 | 4.7 | 8.0 | 110 | 1520 | 21.8 |
| 100 PFC | 8.33 | 100 | 50 | 6.7 | 4.2 | 8.0 | 86.6 | 1060 | 16.7 |
| 75 PFC | 5.92 | 75 | 40 | 6.1 | 3.8 | 8.0 | 62.8 | 754 | 13.7 |

Data Sheet 23 – Electric Motors:

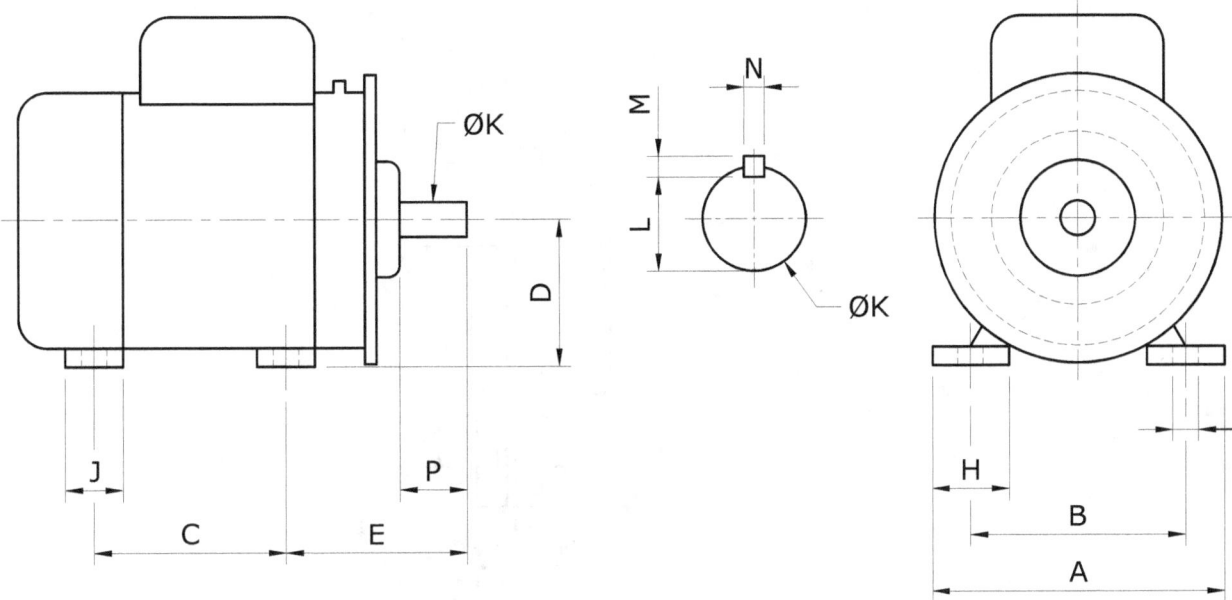

| 1440 RPM | | Dimensions | | | | | | | | | | | |
|---|---|---|---|---|---|---|---|---|---|---|---|---|---|
| 3 Phase | | | | | | | | | | | | | |
| kW | Type | A | B | C | D | E | F | G | H | J | K | L | M x N | P |
| 1.5 | EM-1 | 254 | 203 | 165 | 127 | 156 | 16 | 22 | 67 | 50 | 25 | 21.7 | 6x6 | 58 |
| 2.2 | EM-2 | 254 | 203 | 165 | 127 | 156 | 16 | 22 | 67 | 50 | 25 | 21.7 | 6x6 | 75 |
| 3 | EM-3 | 254 | 203 | 165 | 127 | 156 | 16 | 22 | 67 | 50 | 25 | 21.7 | 6x6 | 75 |
| 3.75 | EM-4 | 280 | 228 | 190 | 140 | 174 | 20 | 22 | 67 | 50 | 28 | 25.2 | 8x11 | 82 |

# MEM09002B – Interpret technical drawing
## Topic 10 – Manufacturer's Catalogues

*Data Sheet 24 – Rigid Shaft Couplings:*

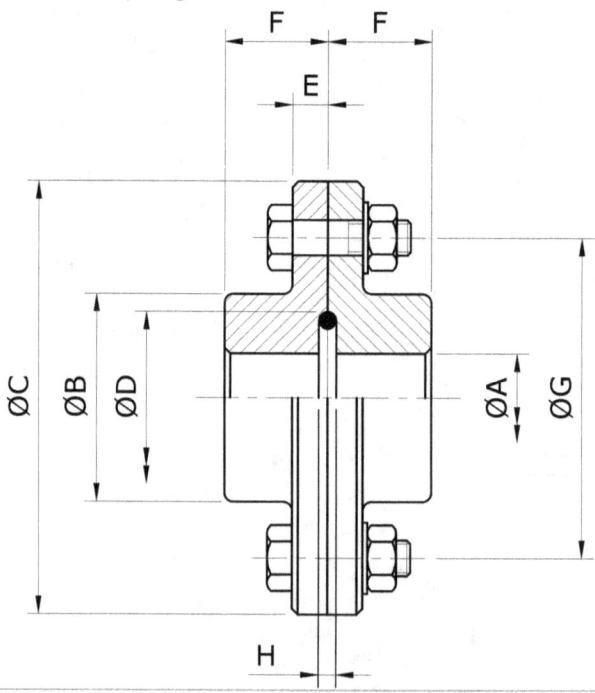

| Type | Dimensions | | | | | | | |
|---|---|---|---|---|---|---|---|---|
| | A (max) | A (min) | B | C | D | E | F | G | H |
| SRC-1 | 42 | 15 | 70 | 148 | 58 | 12 | 56 | 108 | 6 |
| SRC-2 | 48 | 21 | 82 | 171 | 76 | 17 | 61 | 127 | 6 |
| SRC-3 | 58 | 21 | 97 | 198 | 76 | 17 | 68 | 145 | 6 |
| SRC-4 | 70 | 21 | 117 | 216 | 110 | 17 | 76 | 167 | 8 |
| SRC-5 | 78 | 25 | 127 | 254 | 110 | 30 | 68 | 190 | 8 |
| SRC-6 | 85 | 28 | 147 | 279 | 135 | 30 | 100 | 213 | 8 |
| SRC-7 | 105 | 34 | 180 | 330 | 160 | 30 | 117 | 255 | 8 |

**BlackLine Design**
4th October 2015 – Version 3

## Data Sheet 25 – Speed Reducers:

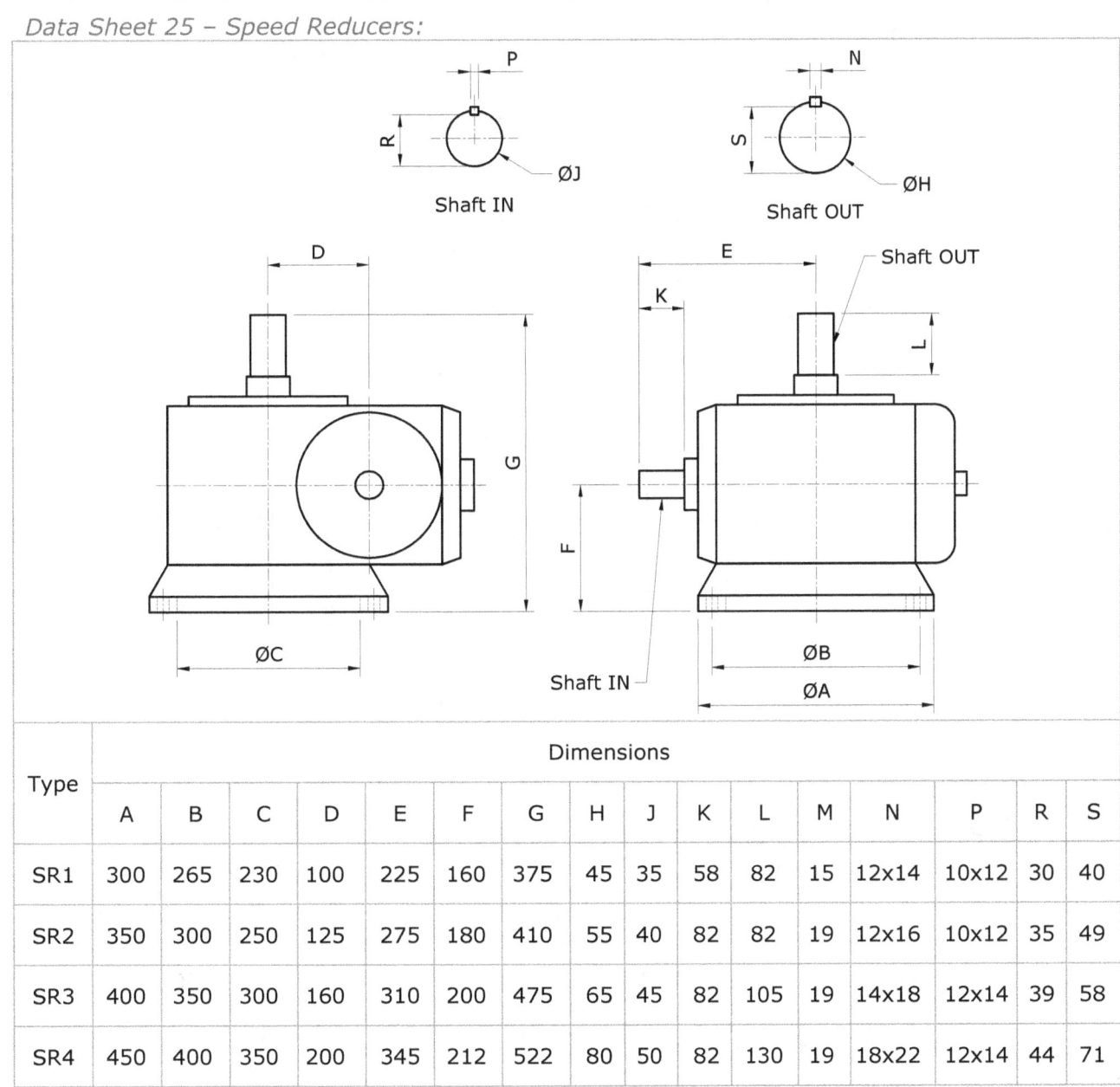

| Type | Dimensions | | | | | | | | | | | | | | |
|------|-----|-----|-----|-----|-----|-----|-----|----|----|----|-----|----|-------|-------|----|----|
|      | A   | B   | C   | D   | E   | F   | G   | H  | J  | K  | L   | M  | N     | P     | R  | S  |
| SR1  | 300 | 265 | 230 | 100 | 225 | 160 | 375 | 45 | 35 | 58 | 82  | 15 | 12x14 | 10x12 | 30 | 40 |
| SR2  | 350 | 300 | 250 | 125 | 275 | 180 | 410 | 55 | 40 | 82 | 82  | 19 | 12x16 | 10x12 | 35 | 49 |
| SR3  | 400 | 350 | 300 | 160 | 310 | 200 | 475 | 65 | 45 | 82 | 105 | 19 | 14x18 | 12x14 | 39 | 58 |
| SR4  | 450 | 400 | 350 | 200 | 345 | 212 | 522 | 80 | 50 | 82 | 130 | 19 | 18x22 | 12x14 | 44 | 71 |

Data Sheet 26 – Flexible Coupling:

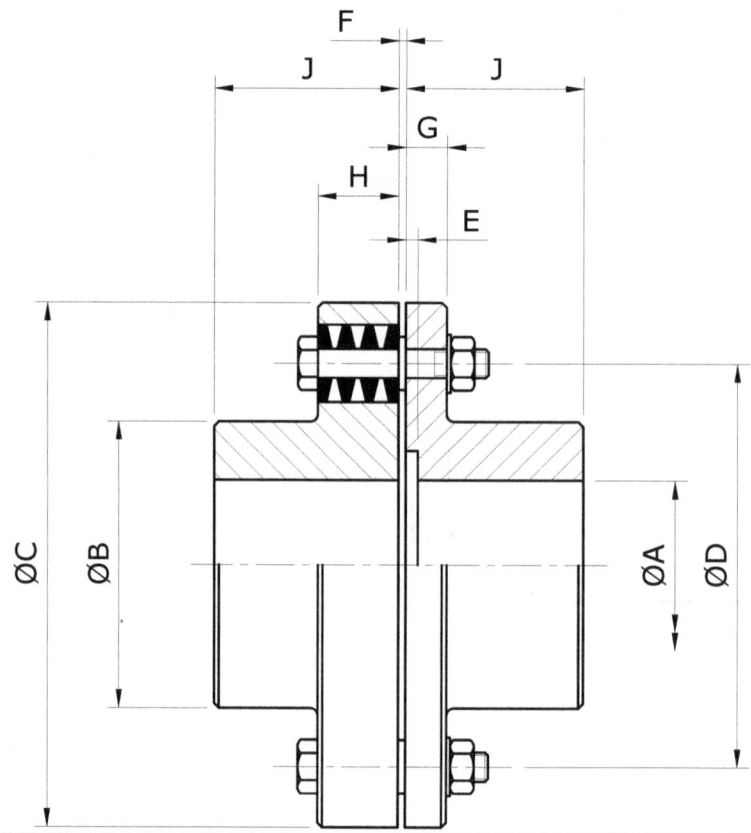

| Type | Dimensions | | | | | | | | | |
|---|---|---|---|---|---|---|---|---|---|---|
| | A (max) | A (min) | B | C | D | E | F | G | H | J | K |
| FC-1 | 70 | 21 | 117 | 216 | 166 | 5 | 3 | 17 | 33 | 79 | 7 |
| FC-2 | 75 | 28 | 127 | 254 | 190 | 5 | 3 | 30 | 56 | 88 | 7 |
| FC-3 | 85 | 28 | 147 | 279 | 213 | 5 | 3 | 30 | 56 | 100 | 7 |
| FC-4 | 105 | 34 | 180 | 330 | 255 | 5 | 3 | 30 | 56 | 117 | 7 |
| FC-5 | 120 | 61 | 206 | 370 | 288 | 8 | 6 | 46 | 76 | 132 | 10 |

MEM09002B – Interpret technical drawing

Topic 10 – Manufacturer's Catalogues

## Skill Practice Exercises:

*Skill Practice Exercise MEM09002-RQ-1001*
Refer to the various Data Sheets, answer the following questions:

1. Is a 10 mm diameter Parallel Spring Pin x 25 mm long available for use in an assembly?

   _____

2. What is the depth for a 360UB44.5 steel beam?

   _____

3. Find the width of a Fi 13 Felt Sealing Ring.

   _____

4. What Bearing number would be required for a Plain Single Groove Deep Ball Bearing 68 mm OD x 40 mm ID and a width of 15 mm?

   _____

5. Select a Locking Nut to suit a thread of M50 x 1.5.

   _____

6. What is the cross sectional area of a 75 x 50 x 8 UA?

   _____

7. What are the centre-to-centre distances of the holes on the feet of the Electric Motor EM-3?

   _____

8. What is the shaft diameter of a Socket Head Cap Screw with a head diameter of 8.5 mm?

   _____

9. What key would be used in a Ø50 shaft?

   _____

10. What diameter groove would be used for a Ø18 Internal Circlip?

    _____

11. What are the Shaft diameter, Outside Diameter and Width of a 72 09 B Single Row Angular Contact Ball Bearing?

    _____

12. What range of shaft diameters can accommodate a 12 mm wide Gibb Head Key?

    _____

13. What Output Shaft diameter is used on a SR2 Speed Reducer?

    _____

Name: _____

# MEM09002B – Interpret technical drawing
## Topic 10 – Manufacturer's Catalogues

14. What is the gauge distance for holes on a 200UC Universal Column?
    _____

15. Is a Ø6 mm diameter x 30 mm long Metric Steel Dowel available for use?
    _____

16. Select a Code Number for an O-Ring to suit a Ø120 mm shaft?
    _____

17. What groove diameter is required for an External Circlip on a Ø63 mm shaft?
    _____

18. What are the Shaft Diameter, Outside Diameter and Width of a TRB-10 Tapered Roller Bearing?
    _____

19. What is the mass of a 50x50x8 EA?
    _____

20. Select a Rigid Shaft Coupling ID number to suit a Ø45 shaft?
    _____

21. What flange thickness is used on a 250UB31.4?
    _____

22. What width groove is required to suit a Ø41 mm shaft?
    _____

23. What are the Shaft Diameter, Outside Diameter and Width of a 2SP-008 Single Row Deep Groove Ball Bearing?
    _____

24. What range of shaft diameters suit a 36x20 key?
    _____

25. What is the nominal diameter of an O-Ring with a code number of 21?
    _____

26. What is the groove diameter to suit a Fi 18 Felt Sealing Ring?
    _____

27. Select a Single Row Angular Contact Bearing to suit a shaft diameter of 65 mm?
    _____

# MEM09002B – Interpret technical drawing
## Topic 10 - Manufacturer's Catalogues

28. What is the width of the tab on a Locking Washer?

29. What is the web thickness of a 460UB81.6 beam?

30. What unequal angle section has a sectional area of 1020 mm²?

31. What Flexible Coupling would be selected if the maximum diameter of the coupling was to be 206 mm?

32. What is the mass/metre, depth and web thickness of a 200 PFC beam?

33. Select the nominal size of a Socket Head Cap Screw with a head diameter of 25 mm?

34. What diameter groove is required to suit a Ø26 bore?

35. Select a Tapered Roller Bearing to suit a maximum bore of 180 mm.

36. What are the Outside Diameter, Width and Thread of a KM 13 Locking Nut?

37. What is the Shaft Diameter and key used for an EM-4 Electric Motor?

38. What is the gap between the coupling parts in a Flexible Coupling FC-1?

39. What is the Outside Diameter and Thickness of a MB 17 Locking Washer?

40. What is the Root Radius of a 460 UB 74.4 steel beam?

MEM09002B – Interpret technical drawing
Topic 8 – Symbols, Notes & Abbreviations

# Practice Competency Test

Referring to drawing MKD-DDg-013 Sheets 1 to 4 answer the following questions.

**Sheet 1:**

1. How many items are required in manufacturing the assembly?

2. What projection has been produced to produce the drawing?

3. What was modified on the drawing to cause a new revision?

4. Give the full name of the material used to manufacture the Jamb Pad?

5. To what standard was the drawing produced?

6. What do the 2 short curved lines represent on the Lever in the End View?

7. If the assembly was to be transported with the Lever in the horizontal position, calculate the total length of the assembly.

8. What is the total number of components that is required for the assembly?

9. Which view does not show the width of the assembly?

10. What is the last step in the assembly before transportation?

Name: _____

MEM09002B – Interpret technical drawing

Practice Competency Test

**Sheet 2:**

11. What type of section is shown?

    _____

12. What is the height of the Ø16 hole above the datum surface?

    _____

13. What is the internal radius of the slots?

    _____

14. What type of view is shown between the Sectioned Front and Top Views?

    _____

15. What does the symbol ⊥ indicate in the End View?

    _____

16. How many tapped holes are required Base?

    _____

17. What is the of line type name for the line connecting the symbols A in the Top View?

    _____

18. What is the meaning of the filled triangular symbol with the letter A attached in the Front View?

    _____

19. What is the overall length of the Base?

    _____

20. Who drew the original drawing and on what date was it drawn?

    _____

Name: _____

MEM09002B – Interpret technical drawing

**Sheet 3:**

21. How many times has the drawing been reissued?

    _____

22. What does the stroke under the R26 dimension indicate?

    _____

23. What do the brackets around the 120 dimension indicate?

    _____

24. What type of section is shown in the Front View?

    _____

25. What is the eccentricity between the R26 outline and Ø16 hole?

    _____

26. What change forced a reissue of the drawing?

    _____

27. What is the name of the firm who produced the drawing?

    _____

28. What is the diameter of the handle of the Lever Arm?

    _____

29. What do the 2-broken lines indicate in the Top View?

    _____

30. What is the size and measurement of the Sheet?

    _____

Name: _____

# MEM09002B – Interpret technical drawing

**Sheet 4:**

31. What orientation is the sheet?

    _____

32. What is the drill size for the M12 tapped hole?

    _____

33. What does the abbreviation C'BORE mean?

    _____

34. What is the thickness of the Splinter Pad?

    _____

35. Who checked the drawing?

    _____

36. What are the centre-to-centre distances between the 3-10.5 holes?

    _____

37. What type of section is shown in the Jamb Pad Details?

    _____

Name: _____

MEM09002B – Interpret technical drawing

*General Questions:*

38. What does P.C.D. mean?

_____

39. Write the abbreviation for assembly?

_____

40. What are the dimensions of an A2 sheet?

_____

41. Arrowheads, Dimension Line, Leader Line, Reference Line, Dimension Text Height and Gap are all features of a dimension; name the other line found in a dimension?

_____

42. From the images below, which represents an isometric drawing?

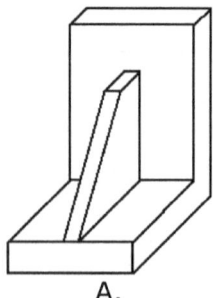 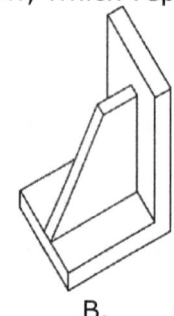 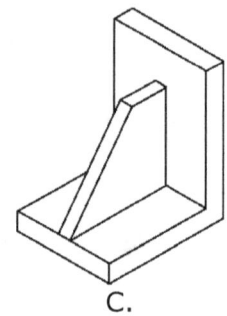 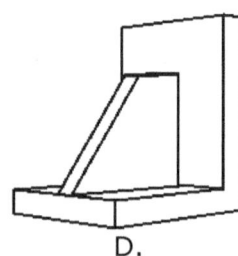

A.　　　　　　　　B.　　　　　　　　C.　　　　　　　　D.

43. What is the abbreviation for EQUIV?

_____

44. From the given Plan View of the Base below, which is the correct Half Section View?

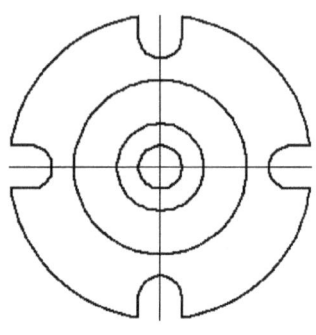 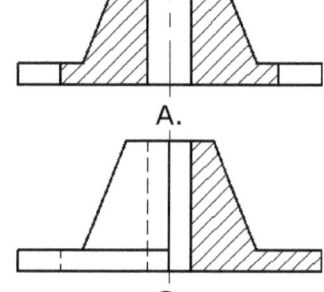 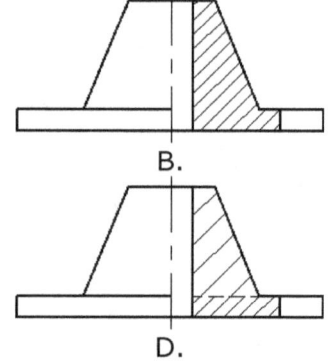

A.　　　　B.

C.　　　　D.

Name: _____

**BlackLine Design**
4th October 2015 – Version 3

MEM09002B – Interpret technical drawing

45. A single thick line around the edge of a sheet is one method of indicating the border; name the other method of border system.

_____

46. What is the purpose of a Revision Block on a drawing?

_____

47. Identify the names of the lettered line styles?

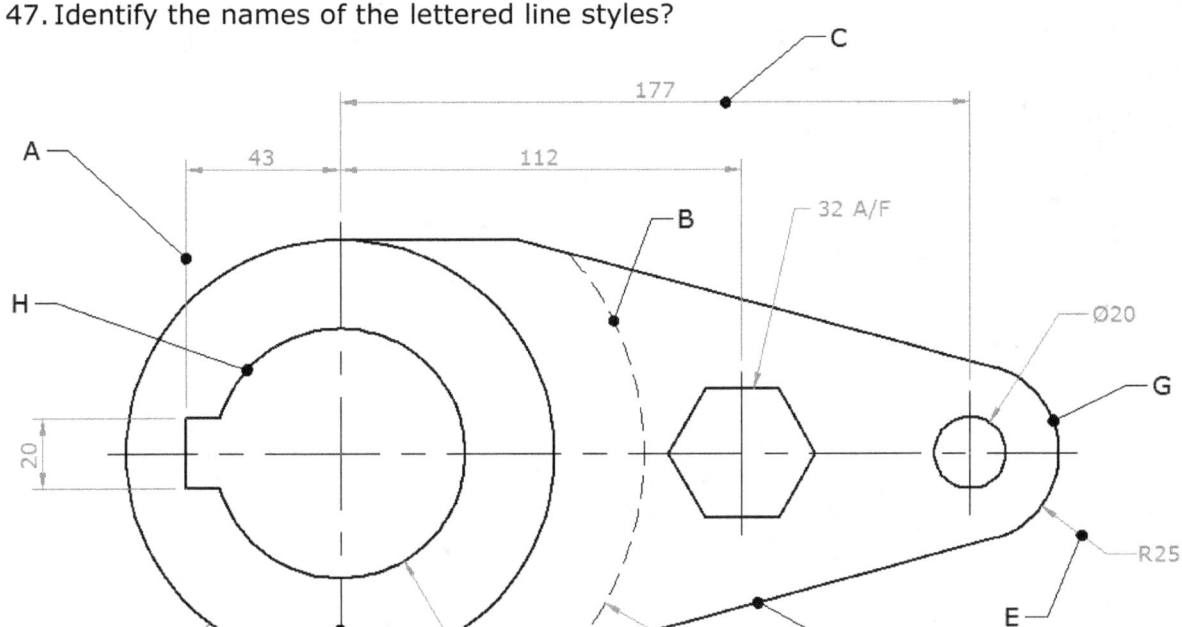

A. _____    B. _____

C. _____    D. _____

E. _____    F. _____

G. _____    H. _____

48. Isometric, Axonometric, Perspective and Oblique are what types of drawings?

_____

49. What is the top and bottom border distances on an A1 sheet according to AS 1100?

_____

50. What is the thickness of a visible outline on A0 sheets?

_____

Name: _____

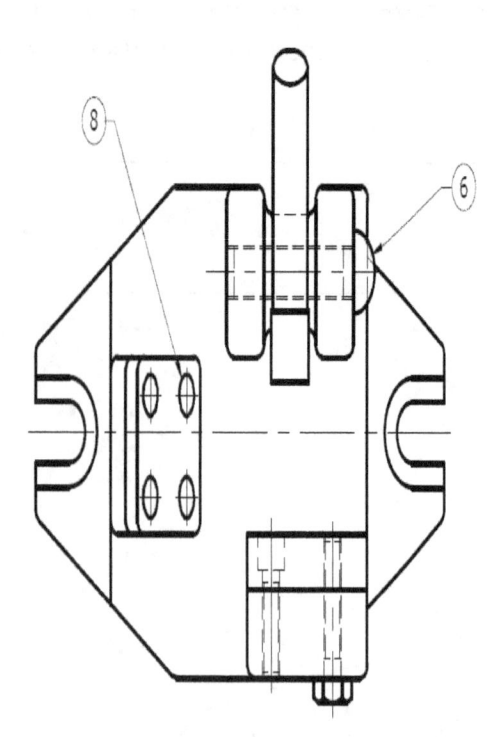
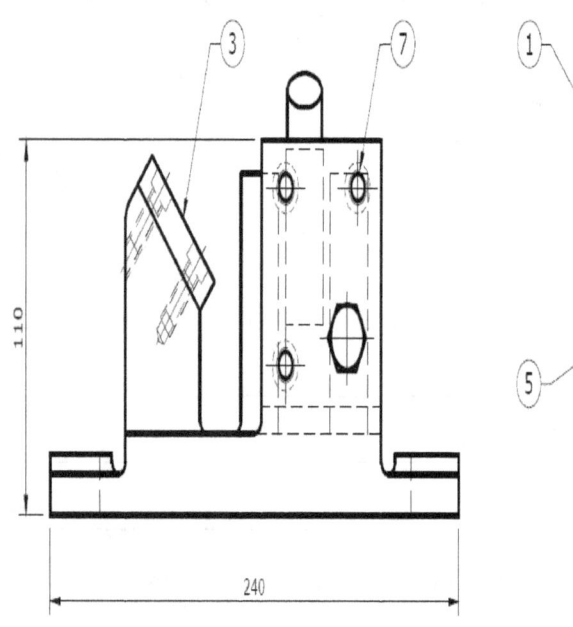
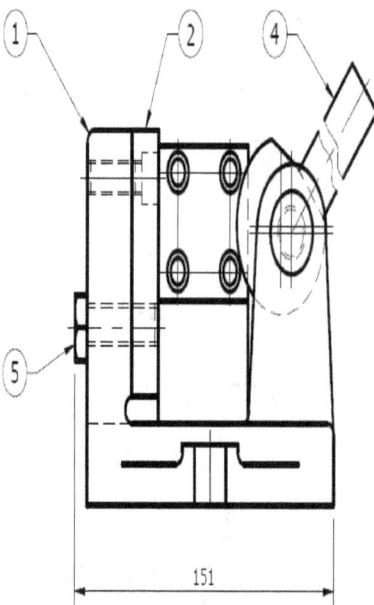

## GENERAL NOTES

1. DRAWN IN ACCORDANCE WITH AS1100 - 1992.
2. REMOVE ALL BURRS AND SHARP EDGES AFTER ASSEMBLY.
3. REFER TO DRAWING SAD-DET-02 & SAD-DET-03 FOR COMPONENT DETAILS.
4. LEVER IS TO HAVE FREE ROTATION OF 150°.
5. FINAL ASSEMBLY TO BE WELL GREASED FOR TRANSPORTATION.
6. ASSEMBLY IS TO FIT CLEARLY INSIDE A PACKING CRATE.
7. ULTRASONIC INSPECTION AFTER ASSEMBLY PER, MIL-STD-2154, TYPE "I", CLASS "II".

### PARTS LIST

| ITEM | DESCRIPTION | QTY | MATL |
|---|---|---|---|
| 8 | SOC HD CAP SCR M8x0.75x25 | 4 | COMM |
| 7 | SOC HD CAP SCR M10x1x30 | 3 | COMM |
| 6 | MACHINE SCR RD HD M16x1.5x70 | 1 | COMM |
| 5 | MACHINE SCR HEX HD M10x1x40 | 1 | COMM |
| 4 | LEVER | 1 | MS |
| 3 | JAMB PAD | 1 | AL AL |
| 2 | SPLINTER PAD | 1 | AL AL |
| 1 | BASE | 1 | CI |

**SCHUMACHER ENGINEERING**

TITLE: STEERING ARM DRILLING JIG ASSEMBLY

| ISU | CHANGE | DRN | DATE |
|---|---|---|---|
| B | NOTE 4 - 150° ROTATION OF LEVER WAS 180° | P.J.H. | 9-6-13 |
| A | ORIGINAL ISSUE | | |

MATERIAL: AS SHOWN
TOLERANCE: ±0.05
SURFACE FINISH: REFER TO DETAILS
DRAWN: S.R.T.
DATE: 3-5-12
CHKD: G.H.W.
A3  SCALE 1:2
DRAWING NO. MKD-DDG-013
SHT. 1  REV. B

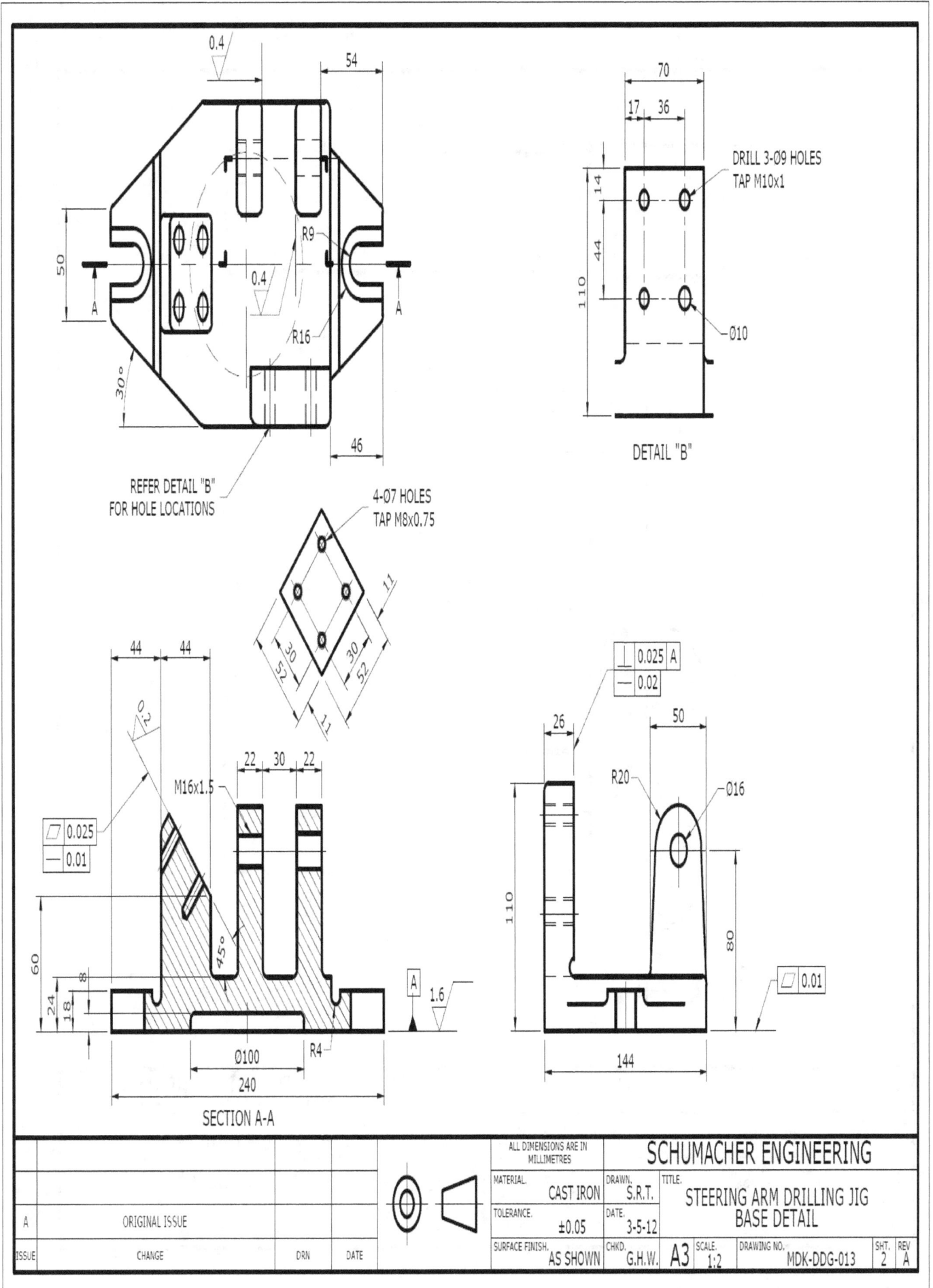

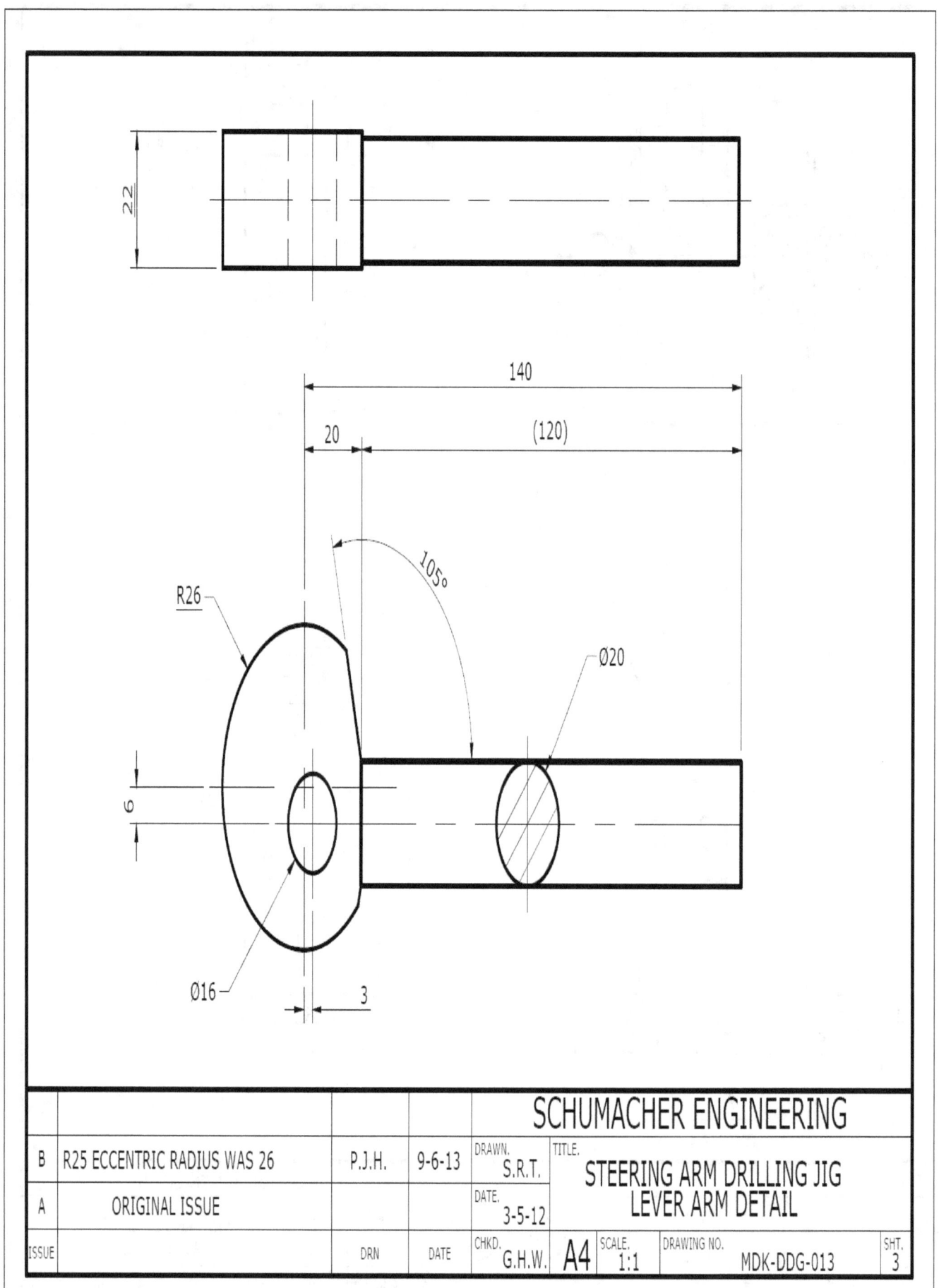

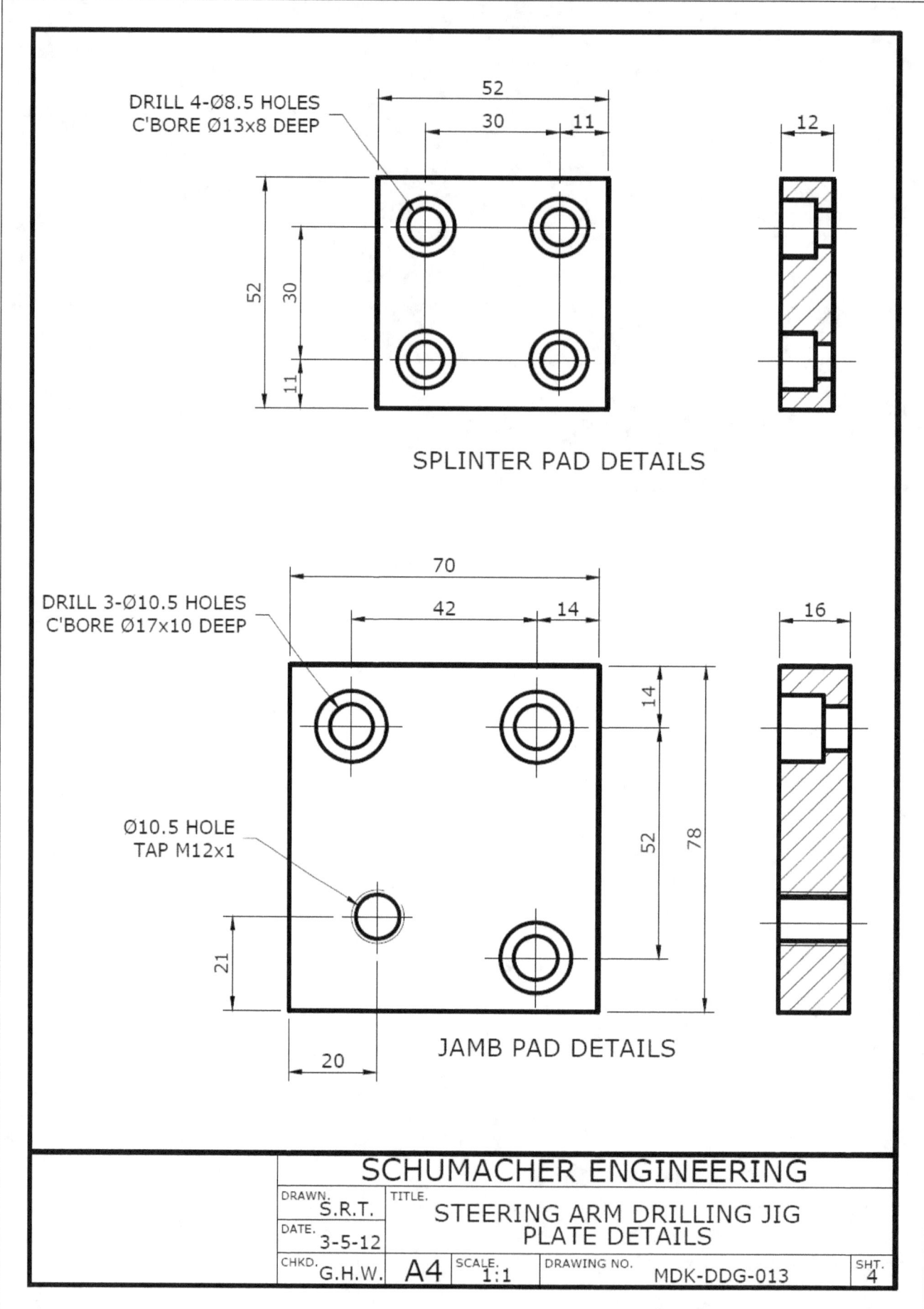

# Appendix:

**Appendix 1 – Abbreviations:**

*Fastenings:*

| | | | |
|---|---|---|---|
| A/F | Across Flats | HEX SOC HD | Hexagonal Socket Head |
| A/P | Across Points | MUSH HD | Mushroom Head |
| CBORE | Counterbore | RD HD | Round Head |
| CH HD | Cheese Head | RSD CSK HD | Raised Countersunk Head |
| CSK | Countersunk | SCR | Screw |
| CSK HD | Countersunk Head | SOC | Socket |
| FILL HD | Fillister Head | SQ HD | Square Head |
| HEX HD | Hexagonal Head | | |

*Materials:*

| | | | |
|---|---|---|---|
| AC | Asbestos Cement | HTS | High Tensile Steel |
| AL | Aluminium | MI | Malleable Iron |
| AL AL | Aluminium Alloy | MS | Mild Steel |
| BRS | Brass | PH BRZ | Phosphorus Bronze |
| BRZ | Bronze | PVA | Polyvinylacetate |
| CAD or CD PL | Cadmium Plate | PTFE | Polytetrafluroethylene |
| CI | Cast Iron | PVC | Polyvinylchloride |
| CS | Cast Steel | RC | Reinforced Concrete |
| CRS | Corrosion Resistant Steel | SS | Stainless Steel |
| GALV | Galvanise | STL | Steel |
| GALV MS | Galvanised Mild Steel | SPR STL | Spring Steel |

*Steel Sections:*

| | | | |
|---|---|---|---|
| L | Rolled Steel Angle | PL | Plate |
| ⊏ | Rolled Steel Channel | RHS | Rectangular Hollow Section |
| CHS | Circular Hollow Section | SHS | Square Hollow Section |
| CRS | Cold Rolled Section | UA | Unequal Angle |
| EA | Equal Angle | UB | Universal Beam |
| HRS | Hot Rolled Section | UC | Universal Column |

*Technical:*

| | | | |
|---|---|---|---|
| ABBR | Abbreviation | DIAG | Diagonal |
| ABS | Absolute | DIAG | Diagram |
| ACCEL | Acceleration | DIA | Diameter |
| AO | Access Opening | ID | Internal Diameter |
| AP | Access Panel | OD | Outer Diameter |
| ACC | Accumulator | DP | Diametral Pitch |
| AR | Acid Resistant | DIM | Dimension |
| AW | Acid Waste | DC | Direct Current |
| ACST | Acoustic | DIST | Distance |
| APC | Acoustic Plaster Ceiling | DWG or DRG | Drawing |
| ADD | Addendum | EA | Each |
| AGGR | Aggregate | EQUIV | Equivalent |
| AIR COND | Air Condition | EST | Estimate |

| | | | |
|---|---|---|---|
| AC | Alternating Current | EXST | Existing |
| AMDT | Amendment | EXT | External |
| ANL | Annealed | FIG | Figure |
| APPROX | Approximate | FLG | Flange |
| ARR or ARRGT | Arrangement | FWD | Forward |
| AS | Australian Standard | FP | Freezing Point |
| ASSY | Assembly | FREQ | Frequency |
| ASSD | Assumed Datum | GA | General Arrangement |
| AHD | Australian Height Datum | GR | Grade |
| AUTO | Automatic | HD | Heavy Duty |
| AUX | Auxiliary | HT or HGT | Height |
| AV or AVG | Average | HEX | Hexagon |
| BRG | Bearing | HP | High Pressure |
| BM | Benchmark | HORIZ | Horizon |
| M | Bending Moment | HYD | Hydraulic |
| BLK | Block | INSUL | Insulation or Insulate |
| BM | Bench Mark | INT | Internal |
| BRKT | Bracket | ID | Internal Diameter |
| HB | Brinell Hardness number | SI | International System of Units |
| BLDG | Building | IP | Intersection Point |
| CALC | Calculated | LAT HT | Latent Heat |
| CAP | Capacity | LMC | Least Material Condition |
| CH | Case Harden | LG | Length |
| CL | Centreline | LEV | Level |
| CG | Centre of Gravity | LIQ | Liquid |
| C/C | Centre-to-Centre | LL | Live Load |
| CHAM | Chamfer | LONG | Longitudinal |
| CIRC | Circle | LP | Low Pressure |
| COL | Column | LUB | Lubricate |
| COMP | Compression | MACH or M/C | Machine |
| CONC | Concentric | MK | Mark |
| CONST | Constant | MATL | Material |
| CST | Crest | MAX | Maximum |
| CYL | Cylinder | MMC | Maximum Material Condition |
| DET | Detail | MECH | Mechanical |
| MP | Melting Point | SHT | Sheet |
| MIN | Minimum | SPEC | Specification |
| MISC | Miscellaneous | SPHER | Spherical |
| MOD | Modification | SPT | Spigot |
| E | Modulus of Elasticity | SF | Spot Face |
| Z | Modulus of Section | SQ | Square |
| I | Moment of Inertia | STD | Standard |
| MTG | Mounting | STA | Station |
| NEG | Negative | STR | Straight |
| NOM | Nominal | SFL | Structural Floor Level |
| NTS | Not To Scale | SW | Switch |
| NO. | Number | SWBD | Switchboard |
| OCT | Octagon | SYM | Symmetry |
| OPP | Opposite | TP | Tangent Point |
| OD | Outside Diameter | TEMP | Temperature |
| OA | Overall | TBM | Temporary Bench Mark |
| PH | Phase | TS | Tensile Strength |
| PCD | Pitch Circle Diameter | THK | Thick |
| PNEU | Pneumatic | THD | Thread |
| POS | Position | TOL | Tolerance |
| POS | Positive | T&G | Tongue & Groove |

| | | | |
|---|---|---|---|
| PRESS | Pressure | XFMR | Transformer |
| PA | Pressure Angle | TRANSV | Transverse |
| QTY | Quantity | TP | True Position |
| R or RAD | Radius | TYP | Typical |
| RECT | Rectangular | ULT | Ultimate |
| RL | Reduced Level | UTS | Ultimate Tensile Strength |
| REF | Reference | UCUT | Undercut |
| RM | Reference Mark | U.N.O. | Unless Noted Otherwise |
| RFS | Regardless of Feature Size | UBP | Universal Bearing Pile |
| REINF | Reinforcement | VAC | Vacuum |
| REQD | Required | VERT | Vertical |
| REV | Revision | VH | Vickers Hardness |
| RH | Right Hand | VOL | Volume |
| RHA | Rockwell Hardness A | WG | Water Gauge |
| RHB | Rockwell Hardness B | WL | Waterline, Water Level |
| RHC | Rockwell Hardness C | WB | Weatherboard |
| $R_a$ | Roughness Value | W | Wide |
| RD | Round | WL | Wind Load |
| SCHED | Schedule | W/O | Without |
| SECT | Section | YP | Yield Point |

**Appendix 2 - Common Engineering Drawing Symbols:**

| Symbol Name | Symbol |
|---|---|
| Datum Identifier | ▲ |
| Diameter | Ø |
| Equal | = |
| Feature Identification | ▭ |
| Projection – First Angle | ◁ ⊕ |
| Projection – Third Angle | ⊕ ◁ |
| Radius | R |
| Slope | ◺ |
| Square | □ |
| Taper | ▷ |
| Counterbore | ⌴ |
| Countersink | ⌵ |
| Depth | ↧ |

## Appendix 3 - Welding Symbols:

### Basic Gas & Arc Welding

| Symbol Name | |
|---|---|
| Fillet | |
| Bead | |
| General Butt | |
| Square Butt | |
| Single Bevel Butt | |
| Single Vee Butt | |
| Single 'U' Butt | |
| Single 'J' Butt | |
| Plug or Slot | |
| Stud | |
| Surfacing | |

### Gas & Arc Supplementary Symbols

| Symbol Name | |
|---|---|
| All Round Weld | |
| Field or On-Site Weld | |
| Backing Strip or Bar | |

### Contour Symbols

| Symbol Name | |
|---|---|
| Flush Weld | |
| Convex Weld | |
| Concave Weld | |

### Resistance Welding

| Symbol Name | |
|---|---|
| Spot | |
| Seam | |
| Mash Seam | |
| Stitch | |
| Mash Stitch | |
| Projection | |
| Flash Butt | |
| Resistance Butt | |

### Resistance Supplementary Symbols

| Symbol Name | |
|---|---|
| All Round Weld | |
| Flush Contour | |

### Symbolic Representation

| Symbol Name | |
|---|---|
| Fillet Located on the Arrow Side | |
| Fillet Located on the Other Side | |
| Fillet Located on Both Sides | |
| Vee Butt Located on the Arrow Side | |
| Vee Butt Located on the Other Side | |
| Vee Butt Located on the Both Sides | |

## Appendix 4 – Structural Steel Sections:

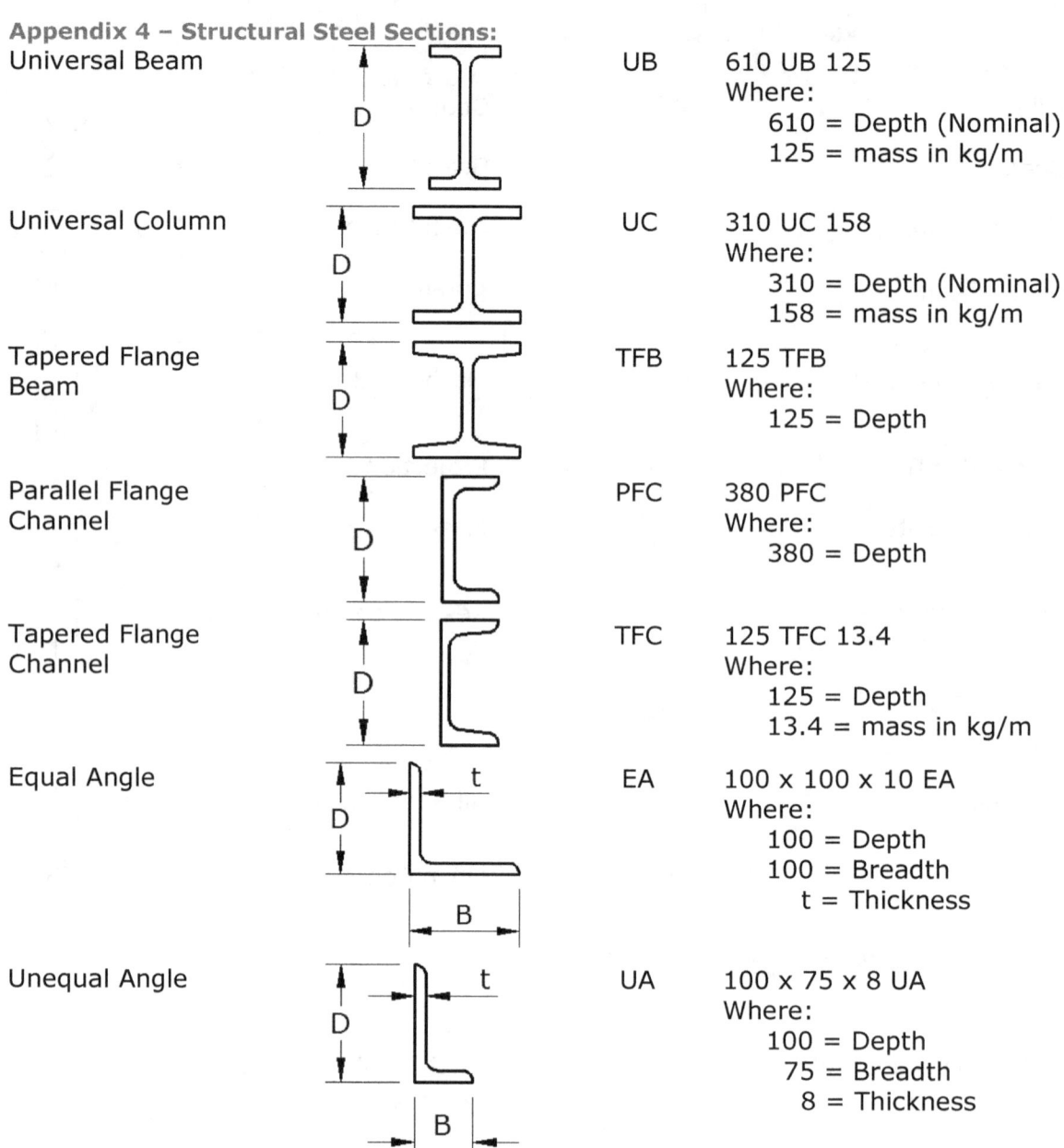

| Section | Code | Example |
|---|---|---|
| Universal Beam | UB | 610 UB 125<br>Where:<br>  610 = Depth (Nominal)<br>  125 = mass in kg/m |
| Universal Column | UC | 310 UC 158<br>Where:<br>  310 = Depth (Nominal)<br>  158 = mass in kg/m |
| Tapered Flange Beam | TFB | 125 TFB<br>Where:<br>  125 = Depth |
| Parallel Flange Channel | PFC | 380 PFC<br>Where:<br>  380 = Depth |
| Tapered Flange Channel | TFC | 125 TFC 13.4<br>Where:<br>  125 = Depth<br>  13.4 = mass in kg/m |
| Equal Angle | EA | 100 x 100 x 10 EA<br>Where:<br>  100 = Depth<br>  100 = Breadth<br>  t = Thickness |
| Unequal Angle | UA | 100 x 75 x 8 UA<br>Where:<br>  100 = Depth<br>  75 = Breadth<br>  8 = Thickness |

*BlackLine Design*
4th October 2015 – Version 3

## Appendix 4 – Structural Steel Profiles:

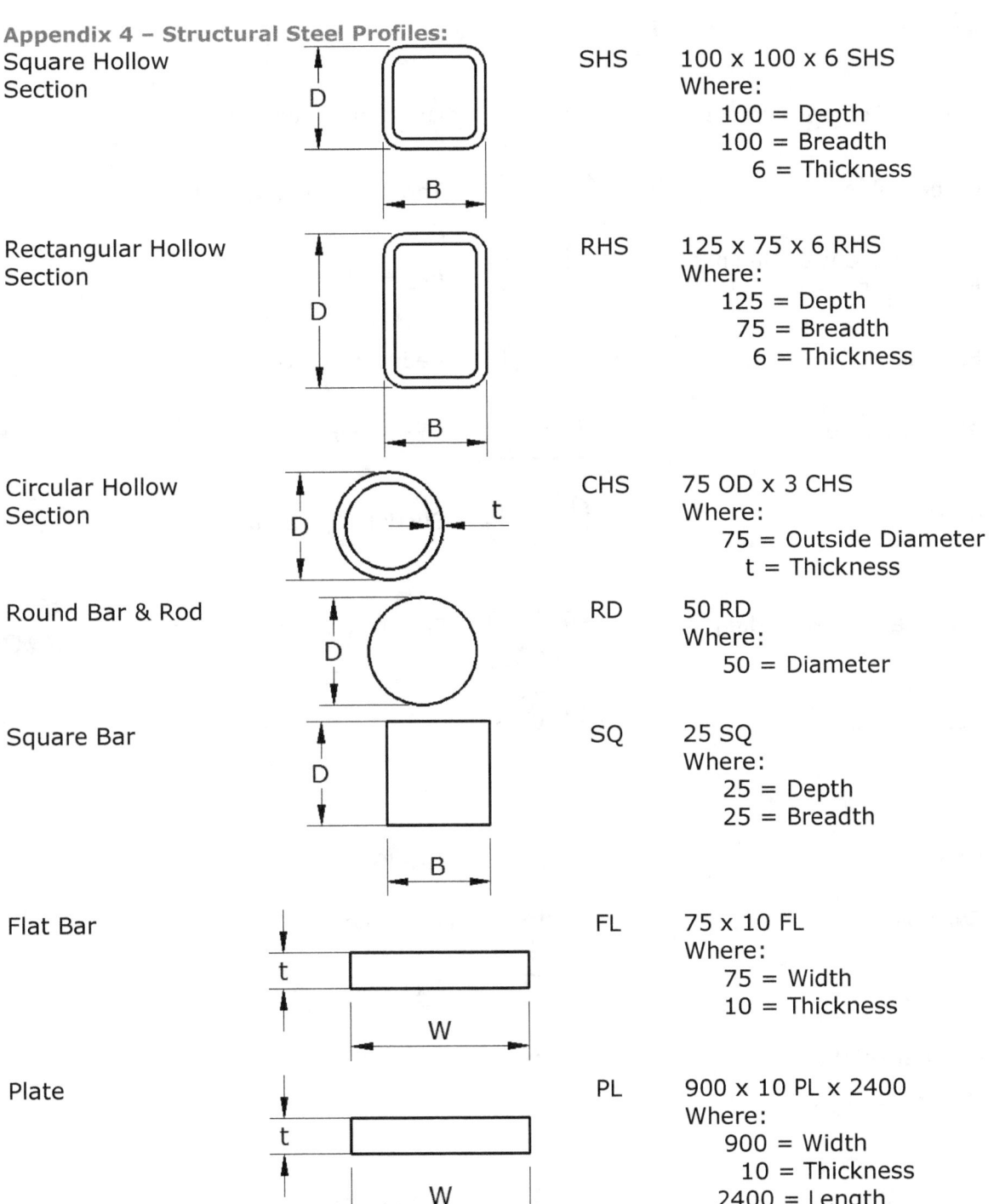

| Profile | | Code | Example |
|---|---|---|---|
| Square Hollow Section | | SHS | 100 x 100 x 6 SHS<br>Where:<br>100 = Depth<br>100 = Breadth<br>6 = Thickness |
| Rectangular Hollow Section | | RHS | 125 x 75 x 6 RHS<br>Where:<br>125 = Depth<br>75 = Breadth<br>6 = Thickness |
| Circular Hollow Section | | CHS | 75 OD x 3 CHS<br>Where:<br>75 = Outside Diameter<br>t = Thickness |
| Round Bar & Rod | | RD | 50 RD<br>Where:<br>50 = Diameter |
| Square Bar | | SQ | 25 SQ<br>Where:<br>25 = Depth<br>25 = Breadth |
| Flat Bar | | FL | 75 x 10 FL<br>Where:<br>75 = Width<br>10 = Thickness |
| Plate | | PL | 900 x 10 PL x 2400<br>Where:<br>900 = Width<br>10 = Thickness<br>2400 = Length |

**Appendix 5 – Pipeline Symbols:**

*Valves:*

| | | | |
|---|---|---|---|
| Valve – General Symbol | ⋈ | Non-Return Valve | ▷ǀ |
| Globe Valve | ⊗ | Three Way Valve | (symbol) |
| Globe Valve with Maximum Flow Adjustment | (symbol) | Angle Valve | (symbol) |
| Ball Valve | (symbol) | Reducing Valve | (symbol) |
| Butterfly Valve | (symbol) | Fire Hydrant | (symbol) |
| Strainer | (symbol) | Suction Pipe Strainer | (symbol) |
| Flow Meter – Recording | R (symbol) | Flow Metre – Non Recording | (symbol) |
| Steam Trap | (symbol) | | |

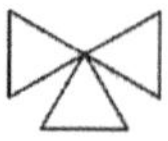

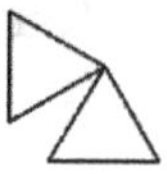

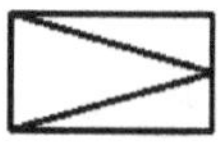

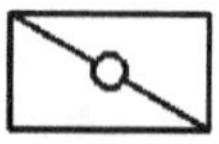

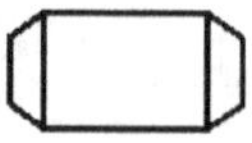

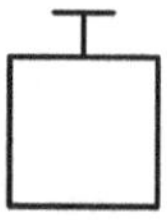

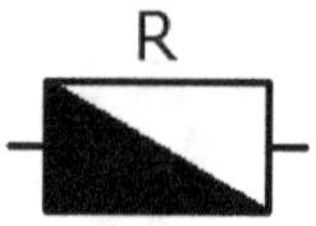

*Operation:*

| | | | |
|---|---|---|---|
| Hand | ⊤ | Diaphragm | (symbol) |
| Solenoid | (symbol) | Lock Shield | (symbol) |
| Spring | (symbol) | Float | (symbol) |
| Counterweight | (symbol) | | |

*Gauges:*

| | | | |
|---|---|---|---|
| Thermostat | [T] | Humidistat | [H] |
| Thermometer Dial | (T) | Pressure Gauge | (P) |
| Sight Glass | (symbol) | | |

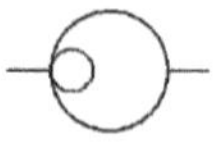

*Miscellaneous:*

| | | | |
|---|---|---|---|
| Centrifugal Pump – Solid Casing | | Centrifugal Pump – Split Casing | |
| Injector or Ejector | | Coil – Heating or Cooling | |
| Pipe Crossing | | Vertical Pipe | |
| Vertical Pipe Rising | | Vertical Pipe Dropping | |
| Direction of Flow | | Rise in the Direction of the Flow | |
| Fall in the Direction of the Flow | | "T" Piece | |
| 90° Elbow | | 45° Elbow | |
| Flanged Connection | | Blank Flange | |
| Union Joint | | Anchor | |
| Hanger | | Orifice Plate | |
| Open Vent | | Anti-Convection Loop | |
| Pipe Guide | | Air Cock | |
| Automatic Air Cock | | Air Vessel | |
| Tundish | | Expansion Bellows | |
| Expansion Bend | | Expansion Bend – Lyre Bend | |
| Strainer – "Y" Type | | Strainer – Line Type | |
| Flexible Pipe (Hose) | | | |

**BlackLine Design**
4th October 2015 – Version 3

**Appendix 6 - Mechanical Symbols:**

*Indication of Surface Texture:*
Surface Texture is indicated on a drawing as shown in the following image:

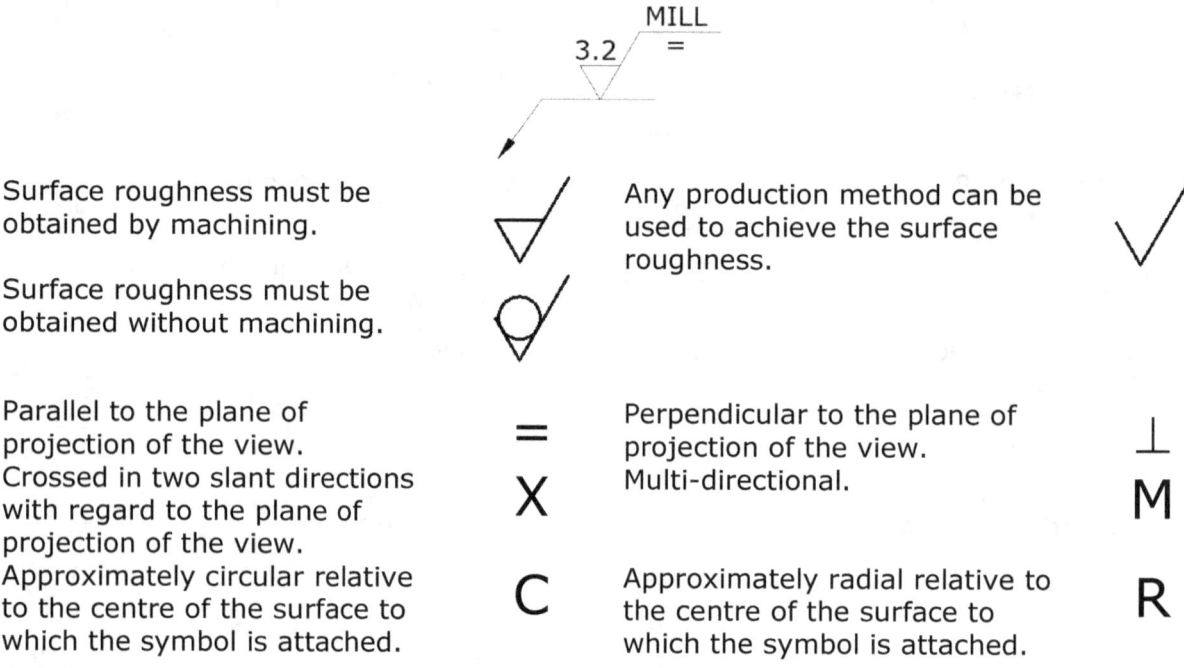

*Geometric Tolerance:*
Geometric Tolerance is applied to the geometry or shape of the component/s through a special symbol shown in the following image:  Typical examples are:
- 2 holes must be concentric to each other.
- A vertical surface must be perpendicular to a datum surface.
- The selected surface must be flat.

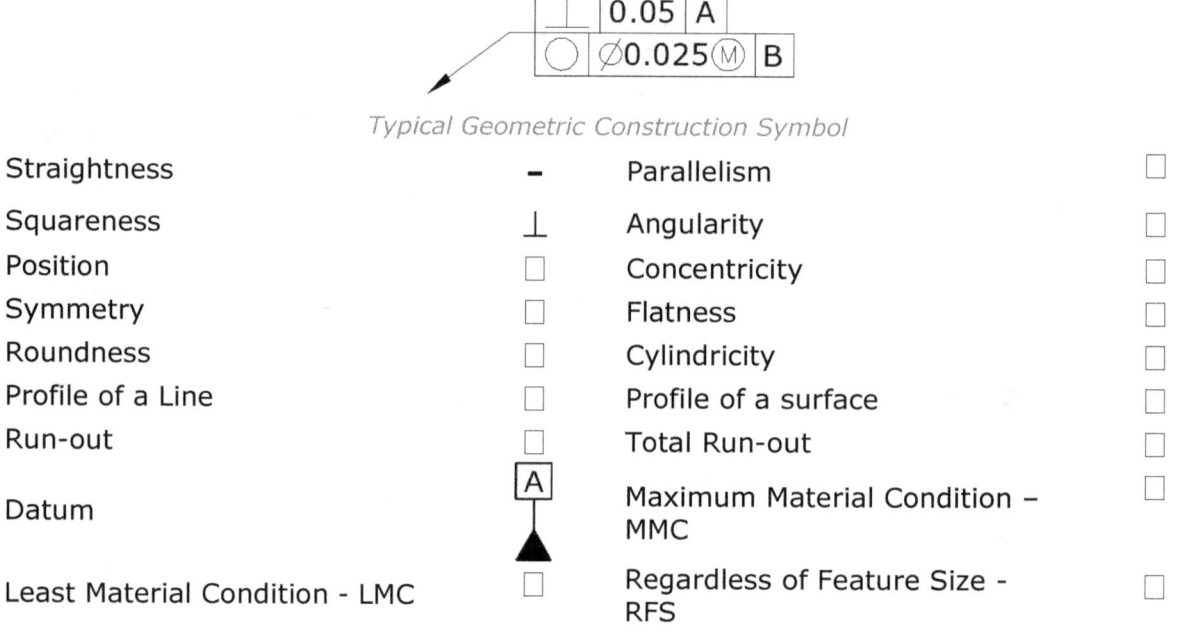

**Appendix 7 – Electrical Symbols:**

*Luminaires & Domestic Appliances:*

| Description | Symbol | Description | Symbol |
|---|---|---|---|
| Luminaire – General Symbol | ○ | Luminaire fixed to wall | ○⏐ |
| Luminaire with number and power of lamps in the group | 5 x 40 W | Luminaire with built-in switch | ○╱ |
| Emergency Lighting Luminaire | ⧖ | Signal Lamp | ⊗ |
| Warning alarm, Panic light | ⊙ | Spotlight | (○→ |
| Floodlight | (○╱ | Lamp with reflector | (○ |
| Fluorescent – Single tube | ⊢──⊣ | Fluorescent – Double tubes | ⊢══⊣ |
| Fluorescent – Triple tubes | ⊢≡⊣ | Fluorescent - Alternate Symbol | 2 x 40 W |
| Discharge Lamp | ⬭ | Auxiliary apparatus for discharging lamp. | ▬ |
| Electrical Appliance – Basic symbol | ▭ | Electric Range | R |
| Garbage Disposal Unit | GD | Exhaust Fan | EF |
| Air Conditioner | AC | Electric Heater | H |
| Fan Heater | FH | Electric Heater – Alternate symbol | ▭▭▭▭ |

*Distribution Boards:*

| Description | Symbol | Description | Symbol |
|---|---|---|---|
| Main Switchboard | MSB | Meter Board | MB |
| Distribution Board | DSB | Automatic Telephone Exchange | PABX |
| Fire Indicator Board | FIB | | |

*Switches & Buttons:*

| Description | Symbol | Description | Symbol |
|---|---|---|---|
| One Way Switches – Single, Two & Three Poles | ⚡ ⚡ ⚡ | Single Pole Pull Switch | ⚡↑ |
| Light Dimmer Switch with variable control | ⚡ | Multi-Position Switch for different degrees of lighting | ⋎ |
| Two-Way Switch | ╱○╲ | Intermediate Switch | ⋈ |
| Time Switch | ⏱ | Push Button | ⊚ |
| Luminous Push Button | ⊗ | Manually Operated Fire Alarm | ▢ |
| Restricted Access Push Button | ▣ | Remote Controlled Equipment | ⌐⌐ |

---

BlackLine Design
4th October 2015 – Version 3

*Socket Outlets:*

| | |
|---|---|
| General Symbol | Multiple Socket Outlet – 'n' for plugs |
| Switched Socket Outlet | Socket Outlet with Protective Earth Contact |
| Single Phase Socket Switched and Earthed | Socket Outlet with Protective Interlocking Switch |
| Multi -Phase Socket Outlet | |

*Miscellaneous:*

| | | | |
|---|---|---|---|
| Point of Attachment | POA → | Earth |  |
| Battery | —| | | |— | Lightning Arrestor | 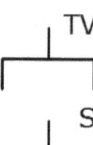 |

*Telecommunications, Radio & Television Apparatus:*

| | | | |
|---|---|---|---|
| General Symbol - Telecommunications | 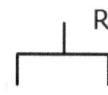 | Television | 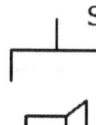 |
| Radio | | Sound |  |
| Aerial – Antenna | 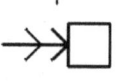 | Loudspeaker |  |
| Radio Receiving Set |  | Amplifying Equipment |  |
| Microphone |  | Telephone Outlet – Wall Mounted | |
| Telephone Installed on Wall | | Telephone Outlet - Floor | |
| Telephone Installed on Floor |  | Intercom |  |
| Through Switchboard |  | Direct Line | |
| Distribution Point | | | |

*Miscellaneous Apparatus and Appliances:*

| | | | |
|---|---|---|---|
| Thermal Fire Alarm Detector Head | ◐ | Motor | Ⓖ |
| Generator | Ⓜ | Ceiling Fan | ⊛ |
| Rectifying Unit DC Power Supply | ▷ | Electric Bell | ⌂ |
| Electric Buzzer | ▽ | Siren | ⇧ |
| Horn | ▭◁ | Clock | 🕒 |

*Cable Codes:*

| | | | |
|---|---|---|---|
| Electric Power | E | Telephony | F |
| Data Circuit | T | Video Circuit | V |
| Audio Circuit | S | Lighting | L |
| Street Lighting | SL | | |

**Appendix 8 – Electronic Symbols:**

*Indicating Instruments:*

| | | | |
|---|---|---|---|
| Ammeter |  | Voltmeter |  |
| Frequency Meter |  | | |

*Contacts for Switches & Relays:*

| | | | |
|---|---|---|---|
| Make Contact |  | Break Contact |  |

*Switchgear:*

| | | | |
|---|---|---|---|
| Circuit Breaker |  | Make Contactor |  |
| Break Contactor |  | Contactor with Coil Type Blow-Out Device |  |

*Coils for Telephone Type Relays:*

| | | | |
|---|---|---|---|
| General Symbol – Relay Coil |  | Relay Coil with 1300 ohm Winding |  |

*Contact Units for Telephone:*

| | | | |
|---|---|---|---|
| Make Contact Unit |  | Break Contact Unit | 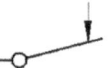 |
| Changeover Contact Unit (break before make) | 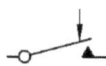 | Changeover Contact Unit (make before break) |  |

*Diode Devices:*

| | | | |
|---|---|---|---|
| General Symbol – Preferred |  | General Symbol – Alternate |  |
| Tunnel Diode |  | Thyristor |  |
| Reverse Blocking Triode Thyristor – 'n' gate, Anode controlled | 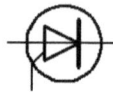 | Reverse Blocking Triode Thyristor – 'p' gate, Cathode controlled |  |
| pnp Transistor (also pnip transistor if omission of the intrinsic region will not result in ambiguity) |  | npn Transistor with collector connected to envelope |  |
| Unijunction Transistor with 'p' type base |  | | |

*Earth and Frame Connections:*

| | | | |
|---|---|---|---|
| General Symbol – Earth or Ground |  | Protective Earth |  |
| Noiseless or Clean Earth Connection |  | Earth Connection | 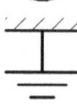 |

*Miscellaneous:*

| | | | |
|---|---|---|---|
| Direct Current or Steady Current – Preferred | ——— | Direct Current or Steady Current – Alternative | − − − − − |
| Alternating Current | ∼ | Conductor or Group of Condensors | ——— |
| Positive Polarity | + | Negative Polarity | − |
| Flexible Conductor | ∿ | Unconnected Cable or Conductor | ⌐ |
| Unconnected Cable or Conductor Especially Insulated | ⌐ | Jumper | − − − − − |
| Two Conductors | // or ═ | Three Conductors | /// or ≡ |
| 'n' Conductors | n | Envelope (Tank) | ◯ |
| Boundary Line | − · − · − | Permanent Magnet | ] |
| Fault | ⚡ | Indicator | ⊖ |
| Hot Cathode – Preferred | | Hot Cathode – Alternate | |
| Photoelectric Cathode | Y | Anode (Plate) or Collector | ⊥ |
| Brush on Slip-Ring | )— | Brush on Communicator | )—⊦ |

## Skill Practice Exercises:

*Skill Practice Exercise MEM09002-RQ-0901*

Refer to drawing STPL-12H-36 and answer the following questions:

1. What grid zone is the Detail of the Oil Grove located?

   _____

2. What dimension and grid zone is the dimension drawn "Not to Scale"?

   _____

3. What type of section is shown in the right side view?

   _____

4. What are the overall dimensions of the Body?

   _____

5. How many surfaces are to be machined?

   _____

# MEM09002B – Interpret technical drawing

## Answers

### Answers:

*MEM09002-RQ-0101:*

| | | | | | |
|---|---|---|---|---|---|
| A. | Isometric | B. | Oblique | C. | One-Point Perspective |
| D. | Isometric | E. | Oblique | F. | Two-Point Perspective |
| G. | Detail | H. | Assembly | J. | Schematic |
| K. | Detail Assembly | L. | Axonometric | M. | Schematic |
| N. | Detail | O. | Assembly | | |

*MEM09002-RQ-0201:*

1. Allows easy location of features on a large drawing
2. 594 x 420
3. 16
4. AS 1612-1999
5. Border, Title Block, Parts/Material/Cutting List, Revision Block, Notes & Legends
6. Lower Right Corner
7. Third Angle Projection
8. Bolt Hex Hd M12x1.5x100

*MEM09002-RQ-0301:*

| | | | | | |
|---|---|---|---|---|---|
| A. | Visible Outline | B. | Hidden Outline | C. | Visible Outline |
| D. | Dimension Line | E. | Centreline | F. | Projection Line |
| G. | Centreline | H. | Hidden Outline | J. | Dimension Line |
| K. | Visible Outline | L. | Projection Line | M. | Leader Line |

*MEM09002-RQ-0302:*

| | | | | | |
|---|---|---|---|---|---|
| A. | Adjacent/Existing Line | B. | Centreline | C. | Visible Outline |
| D. | Leader Line | E. | Adjacent/Existing Line | F. | Break Line |
| G. | Centreline | H. | Hidden Outline | J. | Centreline |
| K. | Pitch Circle Line | L. | Leader Line | M. | Visible Outline |

*MEM09002-RQ-0401:*

1. A dimension is a numerical value expressed in appropriate units of measurement and used to define the size, location, orientation, form or other geometric characteristics of a part.
2. Aligned & Unidirectional Systems.
3. Datum Dimensioning.
4. Dimension Line, Projection Line, Arrow, Dimension Text, Extension & Gap.
5. Projection lines cross the dimension lines.
6. 3 mm x 1 mm
7. The dimension is "Not to Scale".
8. Lines & Symbols, Selection of Distances, Placement of Dimensions, Dimension Standard Features, Precision & Tolerance, and Production Methods.

*MEM09002-RQ-0501:*

| | | | | | |
|---|---|---|---|---|---|
| A. | Third Angle Projection | B. | First Angle Projection | C. | First Angle Projection |
| D. | Third Angle Projection | E. | First Angle Projection | F. | Third Angle Projection |
| G. | None | H. | First Angle Projection | I. | Third Angle Projection |
| J. | Third Angle Projection | K. | Third Angle Projection | L. | First Angle Projection |

M. First Angle Projection    N. Third Angle Projection    O. None

*MEM09002-RQ-0502:*
| | | | | | |
|---|---|---|---|---|---|
| A. | First Angle Projection | B. | Third Angle Projection | C. | Third Angle Projection |
| D. | Third Angle Projection | E. | First Angle Projection | F. | Third Angle Projection |
| G. | Third Angle Projection | H. | Third Angle Projection | I. | None |
| J. | First Angle Projection | K. | Third Angle Projection | L. | Third Angle Projection |
| M. | First Angle Projection | N. | Third Angle Projection | O. | Third Angle Projection |

*MEM09002-RQ-0601:*
| | | | | | |
|---|---|---|---|---|---|
| A. | Offset | B. | Half | C. | Broken |
| D. | Half | E. | Revolved | F. | Aligned |
| G. | Full | H. | Removed | I. | Full |
| J. | | K. | | L. | |
| M. | | N. | | O. | |

*MEM09002-RQ-0602:*
1. B    2. C    3. B    4. A    5. C

*MEM09002-RQ-0701:*
All dimensions are in millimetres unless stated otherwise.

| Scale | A | B | C | D | E | F | G | H | I | J | K | L |
|---|---|---|---|---|---|---|---|---|---|---|---|---|
| 1:1 | 25 | 73.5 | 106.5 | | | | | | | | | |
| 1:10 | 250 | 735 | 1065 | | | | | | | | | |
| 1:100 | 2.5m | 7.35m | 10.65m | | | | | | | | | |
| 1:2 | | | | 10 | 76 | 177 | | | | | | |
| 1:20 | | | | 100 | 760 | 1770 | | | | | | |
| 1:200 | | | | 1m | 7.6m | 17.7m | | | | | | |
| 1:2.5 | | | | | | | 81 | 120 | 216 | | | |
| 1:25 | | | | | | | 810 | 1200 | 2160 | | | |
| 1:250 | | | | | | | 8.1m | 12m | 21.6m | | | |
| 1:5 | | | | | | | | | | 85 | 330 | 588 |
| 1:50 | | | | | | | | | | 850 | 3300 | 5880 |
| 1:500 | | | | | | | | | | 8.5m | 33m | 58.8m |

*MEM09002-RQ-0702:*

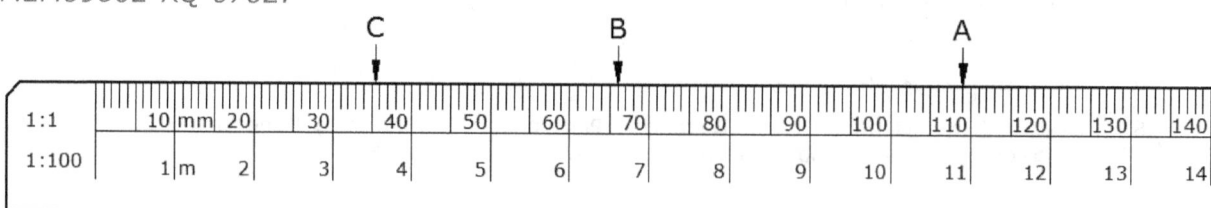

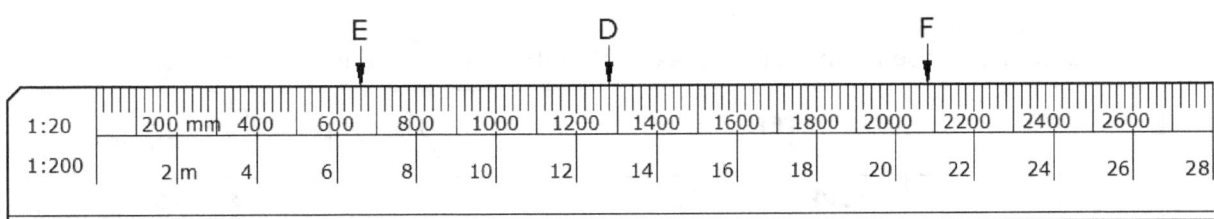

*MEM09002-RQ-0703:*

1. 1455 mm
2. 37.5 mm
3. 1 200 mm
4. 29.4 mm
5. 298 mm
6. 74.25 mm
7. 1 271 mm
8. 5 mm
9. 1 351.6 mm
10. 17 482.5 mm

*MEM09002-RQ-0801:*
The correct response is highlighted in **BOLD** font.

| | | | |
|---|---|---|---|
| E | Modulus of Elasticity | FILL HD | **Fillister Head** |
| **REF** | Reference | SQ | **Square** |
| CRS | **Corrosion Resistant Steel** | **SPEC** | Specification |
| FREQ | **Frequency** | L | Rolled Steel Angle |
| **HTS** | High Tensile Steel | CHS | **Circular Hollow Section** |
| PCD | **Pitch Circle Diameter** | **SOC** | Socket |
| **GALV MS** | Galvanised Mild Steel | HYD | **Hydraulic** |
| BM | **Bench Mark** | **HEX HD** | Hexagonal Head |
| **BRZ** | Bronze | SPR STL | **Spring Steel** |
| UCUT | **Undercut** | **REQD** | Required |
| DIA | **Diameter** | **EQUIV** | Equivalent |
| **MISC** | Miscellaneous | CBORE | **Counterbore** |
| **ADD** | Addendum | AL AL | **Aluminium Alloy** |
| AP | **Access Panel** *(Across Points is A/P)* | I | Modulus of Inertia |
| CL | **Centreline** | A/F | Across Flats |

*MEM09002-RQ-0802:*
The acceptable response for both names and symbols are shown in the highlighted cells.

| Symbol | Name | Symbol | Name |
|---|---|---|---|
| Ø | **DIAMETER** | | Surface must be obtained without machining |
| | **VALVE** | | **SURFACING** |
| | Hanger | | **EARTH** |
| | Feature Identification | | **DIODE** |
| | **VEE BUTT WELD** | | **FAULT** |
| MSB | Main Switchboard | | Loudspeaker |
| | All Round Weld | | **ONE-WAY SWITCH** |
| | **45° ELBOW** | | Circuit Breaker |
| | Float Operated | | **BUTTERFLY VALVE** |
| | Line Type Strainer | | Angle Valve |
| | **MACHINED SURFACE** | | Slope |
| | Datum | | **ANTENNA or AERIAL** |
| 5 x 40 W | **LUMINAIRE WITH 5-40 W BULBS** | | Spot Weld |
| | Push Button | | Counterbore |
| | **2 TUBE FLUORESCENT LIGHT** | | **THIRE ANGLE PROJECTION** |
| | **TAPER** | | **LYRE TYPE EXPANSION BEND** |
| | Spotlight | | Fillet Weld |
| | **VOLTMETER** | | **ON-SITE WELD** |
| | Tunnel Diode | | First Angle Projection |
| | Permanent Magnet | | **SQUARE** |

*MEM09002-RQ-0901:*
1. Grid Zone 5-B

2. Dimension 30; Grid Zone 4-C
3. Full Section
4. 460 Long x 460 Wide x 448 High
5. 6
6. Third Angle Projection
7. Ammonia Compressor Body Detail
8. Ø11 x 20 deep on 98 PCD
9. Pitch Circle Diameter
10. 5 mm
11. 355 mm
12. 210° for 4 hours
13. ±0.05
14. Ø140 mm
15. 195 mm
16. Detail B added
17. R.J.B.
18. STPL-12H-13
19. 1:2
20. Prior to coating
21. ACRU.COM.AU
22. Mild Steel
23. 28 mm
24. 0.05 mm
25. Unless Noted Otherwise
26. Direct or Limit of Size

*MEM09002-RQ-0902:*
1. Mount 52 Hydraulics System
2. 1
3. D
4. 2
5. 2
6. S.C.B.
7. HC145-58
8. J.W.B. 3-12-12
9. 3
10. 3
11. 14
12. Return line deleted from Circuit B
13. 2
14. Not to Scale

*MEM09002-RQ-1001:*
1. No
2. 352 mm
3. 6.5 mm
4. PB-009
5. KM10
6. 921 mm²
7. 203 mm x 165 mm
8. M5
9. 16x10
10. 19 mm
11. Shaft = 45 mm, OD = 85 mm, Width = 19 mm
12. 45 mm – 50 mm
13. Ø55 mm
14. 140 mm
15. Yes
16. 54
17. 60.5 mm
18. Shaft = 50 mm, Outside Diameter = 110 mm, Width = 29.25 mm
19. 5.68 kg/m
20. SRC-2
21. 8.6 mm
22. 1.85 mm
23. Shaft = 35 mm, Outside Diameter = 62 mm, Width = 14 mm
24. 130 mm to 150 mm
25. 29 mm
26. 99 mm
27. 72 13 B
28. 8 mm

29. 9.2 mm
30. 100x75x6 UA
31. FC-5
32. 22.9 kg/m, 200 mm, 6 mm
33. M16
34. 27.2 mm
35. TRB-17
36. Outside Diameter = 65 mm, Width = 12 mm, Thread = M65x2
37. Ø28, 8x11
38. 3 mm
39. Outside Diameter = 119 mm, Thickness = 1.75 mm
40. 11.4 mm

*MEM09002-PT-01:*
### Sheet 1:
1. 8
2. Third Angle Projection
3. The rotation of the Lever was changed from 150° to 180°.
4. Aluminium Alloy
5. AS 1100-1992
6. A break line.
7. 266 mm
8. 13
9. The Front View
10. The assembly must be greased prior to transportation.

### Sheet 2:
11. Offset Section
12. 80 mm
13. 9 mm
14. Auxiliary View
15. Perpendicular
16. 8 (4-M8 +3-M10 + 1-M16)
17. Cutting Plane or Centreline is also acceptable
18. Datum Surface
19. 240 mm
20. S.R.T. on 3-5-12

### Sheet 3:
21. 1
22. The dimension is Not To Scale
23. A Reference Dimension
24. Revolved Section
25. 3 mm & 6 mm
26. The R25 was changed to R26
27. Schumacher Engineering
28. Ø20
29. The Ø16 hole in the Front View
30. A4, 297 mm x 210 mm

### Sheet 4:
31. Portrait
32. Ø10.5
33. Counterbore
34. 12 mm
35. G.H.W.
36. 42 mm & 52 mm
37. Offset Section

### General Questions:
38. Pitch Circle Diameter
39. ASSY
40. 594 mm x 420 mm
41. Projection (or Extension) Line
42. C
43. Equivalent
44. B
45. Grid Zone Border
46. To record the changes to the original drawing.
47. A. Projection Line
    B. Hidden Line
    C. Dimension Line

    D. Visible Outline
    E. Leader Line
    F. Centreline
    G. Visible Outline
    H. Visible Outline
48. Pictorial
49. 14 mm
50. 0.7 mm

www.ingramcontent.com/pod-product-compliance
Lightning Source LLC
Chambersburg PA
CBHW081046170526
45158CB00006B/1879